Mark Anthony Benvenuto
Lunar Chemistry

Also of interest

Industrial Organic Chemistry
2nd Edition
Mark Anthony Benvenuto, 2024
ISBN 978-3-11-132991-8
e-ISBN 978-3-11-133035-8

Industrial Inorganic Chemistry
2nd Edition
Mark Anthony Benvenuto, 2024
ISBN 978-3-11-132944-4
e-ISBN 978-3-11-132951-2

Industrial Chemistry for Advanced Students
2nd Edition
Mark Anthony Benvenuto, 2023
ISBN 978-3-11-077874-8
e-ISBN 978-3-11-077876-2

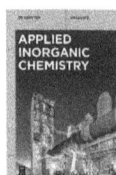

Applied Inorganic Chemistry
Volume 1–3
Rainer Pöttgen, Thomas Jüstel and Cristian A. Strassert (Eds.), 2022
Set-ISBN 978-3-11-074233-6

Mark Anthony Benvenuto

Lunar Chemistry

Lunar Mining, Space Exploration

De Gruyter Textbook

DE GRUYTER

Author
Prof. Dr. Mark Anthony Benvenuto
1202 Marywood Dr.
Royal Oak, MI 48067
USA
benvenma@udmercy.edu

ISBN 978-3-11-138712-3
e-ISBN (PDF) 978-3-11-138730-7
e-ISBN (EPUB) 978-3-11-138788-8

Library of Congress Control Number: 2025938525

Bibliographic information published by the Deutsche Nationalbibliothek
The Deutsche Nationalbibliothek lists this publication in the Deutsche Nationalbibliografie;
detailed bibliographic data are available on the internet at http://dnb.dnb.de.

© 2025 Walter de Gruyter GmbH, Berlin/Boston, Genthiner Straße 13, 10785 Berlin
Cover image: dottedhippo/iStock/Getty Images Plus
Typesetting: Integra Software Services Pvt. Ltd.

www.degruyterbrill.com
Questions about General Product Safety Regulation:
productsafety@degruyterbrill.com

Preface

The colonization of our Moon is still in the realm of science fiction, although several national governments have set their sights on a return of humans to the Moon, and some form of permanent or semi-permanent settlement there. The reasons seem to vary, with the eventual colonization of our neighbor, Mars, being one of them. Whether the distance to get humans safely to Mars can be overcome is still not known, but it is not hard to believe that some brave souls will try to make the journey when a suitable time comes.

This book is an effort to examine the chemistry that will most likely be involved in establishing some human presence on the Moon. It is not an instruction manual for building a colony, nor is it a series of engineering blueprints. It is, however, an overview of all the chemical processes that will be needed as humanity reaches out to try to settle on our most ancient satellite, the body we have been looking up to at night for millennia.

There are always people to thank in an endeavor like this. My editor, Karin Sora, and her team, especially Marie Hammerschmidt, have been an incredible help at every step of this journey. As well, several of my Brother Rats – those who sweated and suffered through a cadetship with me long ago – have been enormously helpful in providing details of their military experiences. Thanks to Dr. Brad Norwood, as well as to Mark Coan and Alan McGreer. And of course, I have to thank my wife Marye, and my sons David and Christian, who have put up with me as I go through the fits all writers seem to have. Thank you all very much.

https://doi.org/10.1515/9783111387307-202

Contents

Chapter 1
Introduction

1.1 Why go to the Moon?

Humankind has looked to the night skies since ancient times, always wondering just what the points of light were that appeared to be looking back at us. The largest source of light in the night was always the Moon, a body that also appears during the day at various times of the year. For most of history, all we could do was look with our eyes, as we did not have the capability to get anywhere near the edges of our atmosphere, much less beyond it. Eventually, telescopic lenses gave us better views. But the chemistry, physics, and engineering capabilities that arose from the end of the Second World War were enough to get a person into orbit in 12 April 1961, when Yuri Alekseyevich Gagarin, a pilot and cosmonaut of the Union of Soviet Socialist Republics, made history by becoming the first person to be launched into space and to orbit the Earth. This astonishing flight was powered by liquid oxygen and kerosene and was brief by today's standards. Yet this is a milestone as big as any other in history, with some comparing it to the taming of fire, or the invention of the wheel in significance.

The Moon has always been the closest of the heavenly bodies to us, and people have studied its phases for millennia. Evidence exists that numerous ancient civilizations had some means by which they could determine the phases of the Moon, and that they knew when new moons and full moons would occur. For example, while there are several solar alignments in Stonehenge, there appear to be positions of the stones that are related to the Moon as well. The Mayan and Aztec calendars also tracked the phases of the Moon. More relevant to current calendars, the Hebrew calendar is sometimes termed "luni-solar" because days are computed based on both the position of the Sun and the phases of the Moon. Indeed, this is how Easter Sunday is still computed each year, and for many years into the future.

As far as what we can see of the Moon, in the West, centuries ago, the light areas of the Moon were dubbed Lunar Highlands, while the dark areas were termed Maria – with a single one being called a Mare. The belief was that the dark areas were seas – the Latin term being *Mare*.

Despite this long span of gazing, we were unable to determine what the surface of the Moon was composed of, beyond noting that some areas were light while others were dark. This changed when the United States landed Apollo 11 on the Moon, and the beginnings of the study of what is called the lunar regolith commenced. Something as straightforward as the composition of the regolith is extremely important, and will be discussed in some detail in this book. This is because it is far too expensive and difficult to lift building materials to the Moon exclusively – rather, regolith will most likely be used to construct any above-ground buildings or other large, permanent structures. Figure 1.1 is a famous NASA photo of what is now called Earthrise. It

https://doi.org/10.1515/9783111387307-001

is obvious that it shows the surface of a portion of the Moon in much more detail than had ever been seen before.

Figure 1.1: Earthrise.
[Figure courtesy of NASA]

The Moon always shows the same "face" to us, and largely because of this, it was thought some pronounced areas must be water, which is why they were given the Latin name "Mare," as mentioned. Thus, there are areas still known as the Sea of Tranquility (Mare Tranquillitatis) or the Sea of Cold (Mare Frigoris), to name just two.

As to why we wish to return to the Moon, there are several reasons that have been given through official channels [1]. Learning more about our own planet, and about the Sun, are two of them. Learning more about the composition of the regolith is another. Additionally, it is believed that what can be learned about humans in a non-Earth environment can be useful as we prepare to launch a manned mission to Mars. At the NASA website concerning Artemis, it states:

> With NASA's Artemis campaign, we are exploring the Moon for scientific discovery, technology advancement, and to learn how to live and work on another world as we prepare for human missions to Mars [1].

In addition to these official reasons for a return to the Moon, a more economic reason – an unofficial one – is that a human presence on the Moon will allow us to determine if there are exploitable materials, such as minerals and helium-3, that can be brought back to Earth. Indeed, taking action to determine the composition of the lunar surface was one of the important first priorities of the Apollo 11 mission. In the archived mission report, it states:

. . . Armstrong went ahead with his scheduled task of collecting the contingency sample – several pounds of lunar surface material which he stowed in a spacesuit pocket. In the course of his collecting, he noted that as he dug down five or six inches below the surface, he encountered hard, cohesive material. Armstrong had been assigned this collection as a first task to make sure that there would be samples aboard in case an early abort of the mission was necessary [2].

It is apparent that one important reason for getting to the Moon at all was to see what the regolith was composed of, and to bring some back to Earth for study.

1.2 What we will do on the Moon?

In the 1960s, the race to the Moon – that between the United States and the Soviet Union – was largely one of national pride, and of two competing economic and political systems. Being able to plant a flag there, and bring back some soil samples, were two of the major objectives. The idea of creating a sustainable human presence on the Moon was still the stuff of science fiction – something we can continue to claim at the present moment. But there are significant differences in technology and experience in near-Earth space travel between that time and today, ones that should help us to establish a long-term human presence on the lunar surface.

It has been said numerous times and repeated here that a presence on the Moon is a step on a longer-term journey to a presence on Mars. But to justify this, it is logical to expect some more immediate benefits than simply gathering data on how human bodies perform on our Moon, and are affected by a lower-than-normal gravity. A greater understanding of our Sun and our own atmosphere are two possibilities, just mentioned, although they may be considered to be of limited practical use [3].

A perhaps obvious example of a more economically based reason for a lunar presence is the mining of various materials, which we will discuss in detail in Chapter 7. Some people will undoubtedly find this ethically offensive, and view the idea of mining the Moon as a desecration of a pristine body in our solar system, one that has been untouched for roughly 4.5 billion years. Others will point out that there are materials on or near the lunar surface that are valuable here on Earth, and that would otherwise never be used. A non-exhaustive list of what we believe to be available on the Moon is shown in Figure 1.2.

Name	Formula	Potential use	Comments
Helium	^3_2He	Medical MRI, plus NMR magnets	Spalled from the Sun
Ilmenite	$FeTiO_3$	Titanium	Valuable metal on Earth
Olivine	$(Mg,Fe)_2SiO_4$	Magnesium	Used in numerous metal alloys
Plagioclase	$CaAl_2Si_2O_8$ to $NaAlSi_3O_8$	Aluminum	A type of feldspar
Pyroxene	$XY(Si,Al)_2O_6$		X = Ca
			Y = Na

Figure 1.2: Materials potentially available on the lunar surface.

At first glance, it may seem odd to include helium with a series of metals. However, it is believed that helium-3, which is ejected from the sun, may exist widely spread across the lunar surface. On Earth, helium-4 is mined with natural gas and used for super-cooling in hospital magnetic resonance instruments (MRI), as well as research high-field nuclear magnetic resonance (NMR) instruments. The helium mined on Earth is ultimately off-gassed because Earth's gravity is not strong enough to retain it in the atmosphere. This means most high-field NMR instruments in universities and research institutions lose a significant amount of their helium approximately twice a year, at which time it must be refilled (with helium recovery systems being both expensive and still in their infancy). Thus, any helium on Earth is essentially a fossil fuel – a material that cannot be replaced – and the ability to obtain more is critical, even if it seems to be coming from an exotic location.

1.3 How many people can be housed on the Moon?

In the 1960s and 1970s, the Apollo missions to the Moon each had two astronauts in the capsule that landed on the Moon, and a third astronaut in the vehicle which stayed in orbit. The idea of two astronauts actually being on the lunar surface was the result of a combination of factors, including each being a part of the safety protocols for the other, and the idea that more work could be accomplished during any lunar visit with two individuals on the surface than with one.

The longest of any of the Apollo missions that landed on the Moon was the Apollo 17 mission, during which the lunar module stayed on the surface for 74 h, 59 min, and 38 s [4]. What is being planned for the current Artemis missions and beyond is a significantly longer presence (probably 20–40 days), although specific durations have not yet been determined [1]. Ultimately, what appears to be planned is a much longer presence on the lunar surface, although for health and other reasons, crews may have to be rotated after a specific amount of time (which has not yet been determined). There are several health-related reasons for rotation off the lunar surface and back to Earth, but two major ones are the long-term effects on the human body of decreased gravity and the potential for bodily harm caused by any prolonged exposure to direct rays of the Sun with no interference from an atmosphere [5–8].

In addition to the amount of time spent on the Moon, the Artemis missions, and those which will be planned based on data gathered from the Artemis missions, will help determine what is required in terms of supplies, materials, and other needs for supporting life there for extended periods of time. The chemistry involved in each of these steps, and the chemistry involved in human activities on the Moon, is what this book intends to examine, although many other people and sources will also create predictive models. For example, as mentioned, the effects of very low gravity on the human body will be closely studied, even though some data already exist based on what has been gleaned from astronauts who have been on the U.S. space station and on the Soviet/Russian station, Mir, which was operational from 1986 to 2001 [5–8].

Several NASA publications have indicated that establishing some longer-term lunar presence is a stepping stone on a pathway that leads to a human presence on Mars. To cynics, this may seem like a "non-reason," but there is a great deal of merit in it. Since the Moon can be described as deep space with some gravity, and Mars can be described as a frozen desert with gravity and a thin atmosphere, the idea of learning how the human body adapts to life on the Moon will indeed give us knowledge of what to expect when missions to Mars move from the realm of science fiction to that of hard science. We have landed unmanned rovers on Mars, and in the process, learned not only a great deal about the Martian surface, but about what happens to the material from which the rovers are made on their journey to our neighboring planet. Having human beings live and work on the Moon will give us data that should then be usable in a more distant future.

References

[1] Artemis – NASA. Website. (Accessed 10 April 2025, as: https://www.nasa.gov/humans-in-space/artemis/).

[2] National Aeronautics and Space Administration Mission Report, Apollo 11. Website. (Accessed 10 April 2025, as: nasa.gov/history/alsj/a11/A11_PAOMissionReport.html).

[3] ESA / Science and Exploration / Human and Robotic Exploration / Orion. Website. (Accessed 10 April 2025, as: https://www.esa.int).

[4] National Aeronautics and Space Administration Apollo 17 Mission Report. Website. (Accessed 10 April 2025, as: https://www.nasa.gov/wp-content/uploads/static/history/alsj/a17/A17_MissionReport.pdf).

[5] Apollo in 50 numbers: Food. Website. (accessed 10 April 2025 as: http:www.bbc.com).

[6] The Human Body in Space – NASA. Website. (Accessed 10 April 2025, as: https://www.nasa.gov/humans-in-space/the-human-body-in-space).

[7] The human body in a microgravity environment: long term adaptations and countermeasures. Website. (Accessed 10 April 2025, as: https://repositorio.pucrs.br/dspace/bitstream/10923/12352/2/).

[8] A. LeBlanc, et al. Muscle Volume, MRI relaxation times (T2) and body composition after spaceflight, *Journal of Applied Physiology*, 2000;89:2158–2164.

Chapter 2
Air

2.1 Where do we get oxygen?

Extracting oxygen from available water

It goes without saying that any people housed on the Moon will need air to breathe on an immediate, constant, never-ending basis. All past missions outside our atmosphere have included tanks of air that are integral parts of the spacecraft, from the first orbital craft to those which eventually landed on the Moon. In getting to the Moon in the future, we will again have to carry some breathable gas mixture in the ships. Even after some permanent or semi-permanent human presence is established in a lunar base, ships traveling to or from the Moon which carry humans will still require air/oxygen tanks. When tanks are loaded on Earth, the matter can be as simple as compressing air, a very well-established process, or making special mixtures [1].

The production of special gas mixes is as established as compressing air and has been practiced by navies throughout the world for decades. For example, diving mixtures for military navies, as well as for commercial fleets, usually involve mixing helium with oxygen. This is so that divers under extreme pressure do not suffer from what is called "the bends," a painful, sometimes fatal condition in which nitrogen in the blood, from normal compressed air, comes out of the body in places other than the lungs. Because of the extremely low solubility of helium in blood sera, diving mixes make the decompression process safe.

Compressed air tanks in general use can be of different designs, but a simple diagram is shown in Figure 2.1. The tank simply has to have a strong steel wall – in a lunar setting, it may be aluminum for decreased weight – with a valve at one end to regulate gas flow. Any lunar setting may require some changes in the basic shape (meaning away from the cylindrical shape that is the current standard), depending on storage space and the environment around the tank. Additionally, double-walled tanks may prove to be the safest, since any puncture in an outer wall would not result in a catastrophic loss of the contained gas.

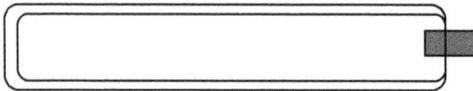

Figure 2.1: Design of an air tank.

Additionally, it may be possible to store oxygen in its liquid form, whether it has been brought from Earth or refined on the Moon, provided containers can be used which

https://doi.org/10.1515/9783111387307-002

keep the oxygen in a cryogenic state. Some military aircraft have what is abbreviated as LOX as a means by which pilots at high altitudes can have breathable oxygen and not suffer from hypoxia. Such apparatuses, when used in military planes, must be small, since all space in a fighter jet or other military aircraft is tightly regulated. It can be imagined that containers of liquid oxygen will have to be larger if they are used in a lunar habitat.

Water has been detected on the lunar surface, although it currently remains uncertain just how much is present. Importantly, water can be split into its elements electrolytically, both being gases [2]. The reaction 2.2 shows this in a simplified form; yet the details of this process bear some examination.

$$2H_2O_{(l)} \rightarrow 2H_{2(g)} + O_{2(g)}$$

Figure 2.2: Oxygen production from water.

This reaction chemistry has been examined in some detail already, but on Earth, a problem with what is called the splitting of water is the cost of electricity. We discuss later in this book the idea of using solar power to produce electricity for any lunar habitation. As long as solar cells or collectors are in sunlight, they should be able to function and produce electricity – which, in turn, can power the splitting of water. This then means the cost of power production becomes connected to the cost involved in building and maintaining solar cells or collectors.

An attractive option for the production of oxygen is the recycling of carbon dioxide. When humans and other animal life exhale, carbon dioxide is expelled, along with a lower amount of oxygen than what is taken in during inhalation. Ideally, after the separation of oxygen from carbon dioxide, the carbon dioxide could, in some way, be chemically split to regenerate the elemental oxygen. The reaction shown in Figure 2.3 is not one that is currently feasible, however.

$$CO_{2(g)} \rightarrow C + O_{2(g)}$$

Figure 2.3: Idealized reclamation of oxygen from carbon dioxide.

Some powerful reducing agent for the carbon, or oxidizing agent for the oxygen, will need to be found so that a driving force can be established to separate the two elements.

Nitrogen

Perhaps obviously, any persons living in a lunar colony will not breathe pure oxygen. They will require a gas such as nitrogen to create a breathable mix similar to that on

Earth. Since there is no known nitrogen on the lunar surface, elemental dinitrogen gas will have to be brought from Earth and mixed with any oxygen produced on the Moon.

Recovering nitrogen, so that it can be in some circular use and re-use in air mixes on the Moon, may require recapture in the form of air liquefaction. Table 2.1 shows how nitrogen is obtained from liquefied air, based on the boiling points of gases, by the slow warming of the liquid. The table indicates the boiling points of the four basic components of air on Earth and shows that cooled oxygen will remain a liquid when nitrogen boils.

Table 2.1: Separation of liquid air by boiling point.

Gas	% in air	Boiling point (°C)
Nitrogen	78	−195.8
Oxygen	21	−183.0
Argon	0.93	−185.8
Carbon dioxide	0.04	−78.5

While this may prove to be intensive in terms of the energy needed to cool any once-breathed gas mixture to the point at which it is liquefied, the process of heating it to separate each component gas is well established. It may become an option in any lunar colony.

Oxygen candles

One method by which oxygen is generated on a submarine, and by which it can also be generated in some lunar habitats, is through what is called an oxygen candle. This is a chemical means by which oxygen gas is produced from a solid and is well-established [3]. Figure 2.4 shows the basic chemistry.

An oxygen candle is a mixture of sodium chlorate and iron powder and is used only as a last resort or a short-term means of producing oxygen should the electrolysis of water on a submarine be in some way disrupted. Each candle is made in a metal shell and weighs roughly 12.5 kg. If the mass of the metal shell can be lessened, this will be advantageous in terms of lifting any oxygen candles to the Moon. The iron powder functions catalytically but is required to keep the chemical reaction going.

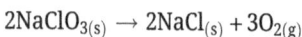

$$2NaClO_{3(s)} \rightarrow 2NaCl_{(s)} + 3O_{2(g)}$$

Figure 2.4: Sodium chlorate decomposition.

As mentioned, iron is mixed directly into the oxygen candle composition but only functions as a catalyst, allowing the candle to burn for a continuous period of time, often hours. Thus, it is not shown in Figure 2.4. Additionally, it can be seen that sodium chloride is the by-product. On submarines, this can be discarded. In any lunar habitat, there may be some use for it that has not yet been foreseen.

As with many materials, all oxygen candles will have to be lifted to a lunar colony. The just-noted mass will have to be accounted for, especially if a large number of them are to be put in place as an emergency source of oxygen. Also, as with their use on submarines, it is expected that should this means of generating oxygen be utilized, it will be as some sort of last resort, used only when the electrolysis apparatus has malfunctioned in some way.

Curiously, sodium chlorate is produced on an industrial scale through the electrolysis of a concentrated brine solution. It is an electrolysis reaction, and the basic reaction chemistry is illustrated in Figure 2.5.

$$NaCl_{(aq)} + 3H_2O_{(l)} \rightarrow elec. \rightarrow NaClO_{3(aq)} + 3H_{2(g)}$$

Figure 2.5: Sodium chlorate production.

If this reaction can be duplicated in any lunar habitat and working environment, it could become a viable second source of oxygen. Evidence already exists for water on the Moon. The presence of sodium chloride, however, has not yet been observed on the Moon.

2.2 How do we regenerate oxygen and air mixes?

The generation of oxygen is key to producing breathable air in any lunar habitat, but human beings do not breathe pure oxygen. As shown, our air is a mixture of 78% elemental nitrogen, 21% oxygen, and the remaining percentage is a mixture of argon, carbon dioxide, and other trace gases. We consider 760 torr to be one atmosphere, but humans who live in mountainous areas tend to acclimate to breathing much lower pressures of air. This may very well have to be taken into account for any personnel living on the Moon.

For sustainable habitation in a lunar base or colony, the nitrogen in any breathable air mixture will be required. At the current time, it seems that this will have to be lifted and imported from Earth.

The possibility exists of utilizing helium in lieu of nitrogen in any breathable gas mixture. As already mentioned, information about the use of helium-oxygen mixtures is well established, largely because divers who descend beyond a certain point in lakes, seas, and oceans must breathe such mixtures. This is to prevent the bends when they surface; as already noted, helium is far less soluble in human blood than nitrogen.

2.3 How can oxygen be harvested from plant growth?

In Figure 2.3, we indicated that some novel chemistry might be required to reclaim oxygen from CO_2. Another method of utilizing carbon dioxide, though, one that will ultimately produce oxygen, is to use such CO_2 in any lunar greenhouses that will be constructed, most likely to grow plants that can be used as a fresh food source, and that will co-produce oxygen gas.

Put very simply, oxygen can be generated in any lunar habitat from plant growth in greenhouses that will have to be constructed on or near the surface. On Earth, the science of greenhouses is well established, with many utilizing soils that are in some way special in order to maximize plant growth. Also, greenhouses can be hydroponic, meaning they can function without soil, using only water and nutrients to grow plants. This will be explored more in Chapter 4, when we discuss food production and storage.

Likewise, vertical farming has begun to take hold in various locations throughout the world; and from it, much can be learned about maximizing the growth of plants in small spaces [5–7]. Organizations that advocate for this type of food production, and companies that build and run vertical farms, perhaps obviously, tout the advantages of it – such as the ability to grow up to 40 times the plant matter per square foot as a traditional farm. While the costs of such farming are not routinely made public, since they can be much higher than those associated with traditional farming, the economics of such enterprises will be initially less important than the ability to produce food locally – and to produce oxygen. Since reactions involving photosynthesis tend to focus on the plant matter produced, it is worthwhile to examine this in terms of the production of oxygen. Figure 2.6 shows an example, starting with 1 ton of carbon dioxide.

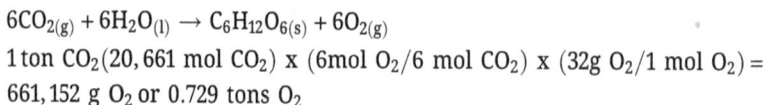

$$6CO_{2(g)} + 6H_2O_{(l)} \rightarrow C_6H_{12}O_{6(s)} + 6O_{2(g)}$$
$1\,ton\ CO_2\,(20,661\ mol\ CO_2)\ x\ (6mol\ O_2/6\ mol\ CO_2)\ x\ (32g\ O_2/1\ mol\ O_2) =$
$661,152\ g\ O_2$ or 0.729 tons O_2

Figure 2.6: Production of oxygen via photosynthesis.

As well as greenhouse types, greenhouse location is important on Earth – and will be on the Moon – in terms of maximizing sunlight exposure for growing plants. In the United States, what can be considered commercial greenhouses are most prevalent in several states with winters that would otherwise not permit plant growth, at least in certain areas. Table 2.2 is a non-exhaustive list of these.

The United States certainly is not the only nation utilizing greenhouses for horticulture in some way. The Netherlands has seen extensive use of greenhouses, with the city of Westland having so many that it is now known as the Glass City – because of the many glass-walled greenhouses. Similarly, greenhouses are used extensively in Iceland, which would otherwise only be able to grow potatoes and root vegetables on

Table 2.2: Greenhouses in the United States [4].

State	No.	Major crops	Comments
California	427	Tomatoes, lettuce, eggplants, herbs	Northern CA can have severe winters
Maine	386	Strawberries, spinach, onions	Routinely has severe winters
Michigan	341	Carrots, tomatoes	
New York	435	Tomatoes, cucumbers	
Pennsylvania	593	Lettuce, tomatoes, eggplants, cucumbers	Can have severe winters and late spring snows

the islands. As with the United States, greenhouses in such nations are capable of producing crops throughout what can be very cold, dark winters. The shortness of winter days in the Netherlands and Iceland does mean that more energy input is required so the plants can undergo optimal photosynthesis. One source, *Nature Rising*, indicates: "At optimal growing conditions, each acre in the greenhouse yields as much lettuce as 10 acres outdoors" [6].

Other greenhouses are hydroponic, meaning the plants are grown with no soil but rather in nutrient-rich water solutions. This has proven to be economically feasible for the luxury dining market, in which the appearance of crops such as lettuce and other vegetables must be excellent when brought to the table. Hydroponic greenhouses tend to be extremely strict about any sort of insect presence or other problematic conditions that would lessen the appearance of the final crop. Perhaps obviously, there is expected to be no pest problem in any lunar greenhouses.

The basic chemical reaction for photosynthesis can be represented as follows in Figure 2.7: This is what essentially all terrestrial plants utilize when growing.

$$6CO_{2(g)} + 6H_2O_{(l)} \longrightarrow C_6H_{12}O_{6(s)} + 6O_{2(g)}$$

Figure 2.7: Photosynthesis.

As shown in Figure 2.6, this is the major means by which free, elemental oxygen is produced on Earth as a by-product of the basic material of plant growth. Because horticulturalists tend to study plant growth and the resulting vegetable products intensely, the study of the production of oxygen tends to be a somewhat smaller focus, simply because it is assumed that oxygen is always present.

It appears that growing plants hydroponically, which will be discussed in more detail in Chapter 4, can be an efficient way to generate oxygen. If any local areas on the lunar surface are composed of regolith that can support and enhance plant growth, it may be viable to construct greenhouses which use that regolith and have maximum exposure to sunlight. Advantages would most likely be led by the use of less water than a hydroponic farming system and the utilization of some nutrients from the regolith, assuming it is more than simply silicon oxides.

2.4 How will by-product hydrogen be used?

Should sufficient water be found in the polar regions of the Moon, it can be split electrolytically into its component elements. Figure 2.8 shows the basic chemistry for this in simplified form, a reaction in which hydrogen is reduced and oxygen is oxidized.

$$2H_2O_{(l)} \rightarrow 2H_{2(g)} + O_{2(g)}$$

Figure 2.8: Electrolysis of water.

Uses will have to be developed for the hydrogen extracted from water – the major by-product of the desired oxygen production.

As already mentioned, when water is split into its two component elements using electricity, both are formed as gases. The equation: $2H_2O_{(g)} \rightarrow O_{2(g)} + 2H_{2(g)}$ is not used on a large scale on Earth as a means of generating either material, largely because oxygen can be concentrated through the distillation of liquefied air in a more economically feasible manner, and hydrogen can be generated via the lean-environment stripping of small hydrocarbons, such as methane. The source of this methane is ultimately crude oil or natural gas. In a lunar environment, the energy needed to split water can, in theory, be that captured by some means of solar radiation. This, then, will result in the generation of needed oxygen, but also of a significant amount of hydrogen gas. As mentioned, it will need to be utilized in some way, possibly as a fuel.

On Earth, hydrogen used as fuel is combined with oxygen. The reaction, which is the reverse of 2.4 – both an addition reaction and a redox reaction – forms the basis of this use. The fact that it is a "circular" reaction, meaning that first the water is separated and then it is reformed, is what limits the two reactions from being economically feasible. Essentially, the energy that must be put into such a cycle is a problem that cannot be solved. In any lunar environment, this problem may be solved because the energy required can be gleaned from sunlight – an inexhaustible commodity on the lunar surface, assuming operational sites are built where direct sunshine is constant.

References

[1]   D.E. Canfield. Oxygen, a Four Billion Year History, DeGruyter, 2025.

[2]   M. Elzagheid. Water Chemistry, Analysis and Treatment, DeGruyter, 2024. https:////doi.org/10.1515/9783111332468-002.

[3]   W.H. Schechter, R.R. Miller, R.M. Bovard, C.B. Jackson, and J.R. Pappemheimer. Chlorate candles as a source of oxygen, *Industrial and Engineering Chemistry*, 1950;42(11):2348–2353. https://doi.org/10.1021/ie50491a045.

[4]   North America Commercial Greenhouse Market Report 2022. Rising Populations and Growing Demand for Food Are Creating Greater Opportunities for Alternate Farming Methods – ResearchAndMarkets.com. (Accessed 10 April 2025, as: https://www.businesswire.com/news/home/20220819005208).

[5]   Association for Vertical Framing. Website. (Accessed 10 April 2025, as: https://vertical-farming.net).

[6]   USVFA. U.S. Vertical Framing Association. Website. (Accessed 10 April 2025, as: https://www.usvfa.org).

[7]   Nature Rising. (Accessed 10 April 2025 as: https://www.naturerising.ie/the-dutch-horticulture-industry/).

Chapter 3
Water

3.1 How can lunar water be mined, refined, and purified?

After air, water is probably the single most important commodity to have and be able to harness and use for an extended presence of humans in any lunar habitat. Throughout long periods of history, it was believed that the dark areas of the Moon were oceans, hence their names being "Mare," the Latin word for sea, although this has long since been found not to be the case.

More recently, but still for decades, it was thought that the Moon was entirely without water because the temperature of any part of the surface exposed to the Sun was hot enough that water would have boiled off. But more recently still, water has been detected at both of the polar regions, in areas of continuous shadow, buried under a significant amount of regolith [1–15]. It also appears to be disbursed throughout some of the surface regolith formations [3, 8]. Estimates as to how much water is present vary widely, simply because there is not yet a significant body of measurements able to estimate or determine quantities [12, 14].

As with virtually all of the commodities we will discuss, there is a cost involved in lifting water from Earth to the Moon. The general belief is that using water that already exists on the Moon [1, 2] for some aspect of extended living there will be vastly less expensive than launching water from Earth to the Moon in any form of spacecraft.

It is prudent to compare water use by people on Earth with expected water use in a lunar habitat, but this presents certain problems. For example, the use of water by the average resident of Tucson, Arizona is much different from that of a person living in Philadelphia, Pennsylvania, simply because of the proximity of each municipality to some natural body of water [16, 17]. Table 3.1 illustrates water use in Philadelphia, which can serve as an example of water use in a developed city.

Table 3.1: Residential water use in Philadelphia, Pennsylvania [17–22].

Use	Amount (gal)	Times/day	Total (gal.)
Bathing	36	1	36
Dishwasher	6 (est)	<1	6
Drinking	0.125	8	1
Hygiene	2.5	2	5
Shower	20	1	20
Toilet	3	6	18
Washing machine	15	<1	15
			Total, 101

https://doi.org/10.1515/9783111387307-003

These numbers and estimates most likely represent volumes that are far higher than what will be available in a lunar habitat and colony. For example, bathtubs will be a luxury and may be replaced almost entirely by showers with drains that can catch water for recycling. Water use in toilets can be minimized, much as it is on airplanes today. The amount of time spent with running water in showers can also be minimized, as it currently is on military submarines, where submariners must live for weeks or even months at a time (water in this situation is turned on to wet a person, turned off while they soap up, then turned on to rinse). And in virtually all uses, water can be re-claimed and re-used.

Water found on the Moon will be ice, which means its mining will be much like mining any other soft mineral. A variety of means can be used to melt it, all of which require some form of energy input. The easiest may be placing the ice into sealed chambers that are exposed to sunlight.

The purification of water is well established on Earth, and, again, more than one means of purification can be replicated on the Moon. One method that is straightforward is flash distillation, in which water is heated and then condensed. The water that condenses is extremely pure – pure enough for consumption – and any brackish or mineral-laden water can be used in some other manner, possibly for plant growth or plumbing. Figure 3.1 shows the basic scheme of flash distillation.

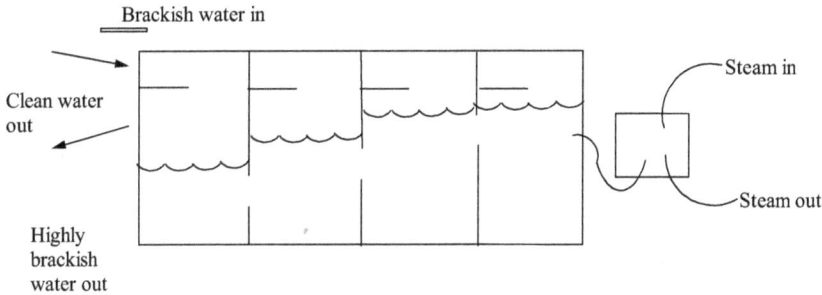

Figure 3.1: Schematic of flash distillation.

Existing flash distillation apparatus operations on Earth never separate brackish water into clean water and solid salt. Rather, a more concentrated brackish water is one of the two final products. While this may still have to be the case in any water purification operation on the lunar surface, one can imagine that even the highly brackish concentrate will be valuable enough to find use in some other application. It is not hard to conceive of such being utilized in building materials, such as in the creation of cement.

3.2 How can water be split, what capabilities and limitations exist on Earth?

In any lunar colony, water will not only be used for daily living – drinking, cooking, and bathing – but will also serve as a starting material for two other commodities that will be of use: hydrogen and oxygen, assuming large enough amounts of it can be found. The simple equation for what is commonly called water splitting is shown in Figure 3.2 and was previously presented in Chapter 2; it represents how water can be divided into its component elements using electrolysis.

$$2H_2O_{(l)} \rightarrow 2\,H_{2(g)} + O_{2(g)}$$

Figure 3.2: Splitting water.

This reaction is one that seems very simple and can be made to occur in a few different ways on Earth, but never without some cost in energy and never at 100% efficiency [23]. This cost has made the large-scale reduction of water to elemental hydrogen and oxidation of it to elemental oxygen prohibitively expensive. However, the same chemistry may be far easier to undertake on the Moon because of the potential power source or sources available there – solar energy.

Quantities of water that can be split are important not only for breathing but also for propulsion. Existing rockets from Earth utilize liquid oxygen as part of their propulsion, in a mix often called kerolox. The name comes from the mixture of kerosene and oxygen. Additionally, rockets can use hydrolox, a mixture of hydrogen and oxygen. Perhaps obviously, it will be easier to produce such a fuel on the lunar surface than to bring it from Earth for later use.

Solar exposure is something that the lunar surface has in abundance, at least for each lunar day. With no atmosphere, sunlight hitting the surface can easily be converted to energy through either the use of solar panels or through concentrated solar power (CSP) using some fluid to transfer the resulting heat. In theory, this can be harnessed for the power needed to split water.

3.3 Can solar panels or concentrated solar power (CSP) be used for water splitting?

The possibility of utilizing some form of concentrated solar power on the Moon is one that may involve importing less equipment and material from Earth than would be needed if solar panels were used for power generation.

Both traditional solar power and the somewhat newer CSP may be efficient ways to convert solar power into electricity in some lunar habitat environments. Both power sources will be discussed in more detail in Chapter 9, but at this point, we can note that by whatever means electricity is produced, it can be used for the electrolytic splitting of water, as shown in Figure 3.2.

3.4 How can water and waste-water be re-used?

The value of water in any lunar colony is as great as, or greater than, the value of water in the driest desert climates on Earth. Thus, a great deal of emphasis will have to be placed on water reclamation and recovery in any lunar habitat.

Currently, on Earth, the means by which water is made drinkable on a large scale can be ranked in terms of cost. These costs escalate as follows, through several categories within these three broader ideas:

First: Distribution efficiency

Virtually all existing water distribution systems in any large area, from rural homes to city apartments and single houses, have the potential for leakage between the source and the end user. Minimizing leaks in the distribution of potable water, through constant monitoring and maintenance, remains the least expensive way to improve the distribution and use of water. Some water distribution systems in urban areas – usually older, well-established ones – are very inefficient, with significant amounts of water being lost to leaks on a continuous basis. In such cases, it is often political will that ultimately gets such problems repaired.

Perhaps obviously, a high level of water distribution efficiency will be critical in any settlement on the Moon. This will involve constant monitoring as a form of routine maintenance. Additionally, this will most likely entail some sort of rapid leak repair kit to stop any leak caused by an accident or one that has sprung unexpectedly. Repair kits involving sleeves or sheaths and hose clamps are an established aspect of plumbing and are used extensively in a wide variety of applications today.

Second: Reclamation

Water that has been used can routinely be cleaned to a level where it can be consumed by people with no ill effects. Often called gray water, such water undergoes several steps in its re-purification, which include:
a. Microfiltration through porous membranes of approximately 0.6 microns, or 0.0006 mm, which removes solid waste.
b. Reverse osmosis, for viral and bacterial decontamination. This requires osmotic membranes, which need to be replaced or cleaned on a regular basis.
c. Irradiation by ultraviolet light. This further disinfects the reclaimed water.

Figure 3.3 illustrates the principle of reverse osmosis. The heart of any reverse osmosis apparatus is the membrane filter that traps ions on one side, while allowing water to pass through. Use of the membranes means that a certain amount of stress is put on them, and they must be replaced on a regular basis.

Most reverse osmosis membranes are made using what is called a thin-film composite (TFC). This often consists of three component layers:

1. A polyamide layer. This allows water to pass but not ions of dissolved salts.
2. Polysulfone layer. This (aryl-unit – SO_2 – aryl-unit) is used predominantly as a support.
3. Polyester support. The durability of polyester is useful in extending the life span of any membrane.

Reverse osmosis membranes can also be made from cellulose acetate, which is highly hydrophilic. Figure 3.4 shows the basic repeating monomer unit of cellulose acetate. The hydrophilic nature of cellulose acetate allows water to pass through a membrane composed of it while impeding the movement of ions through it. These membranes are effective but do degrade somewhat faster than the more widely used TFC composites.

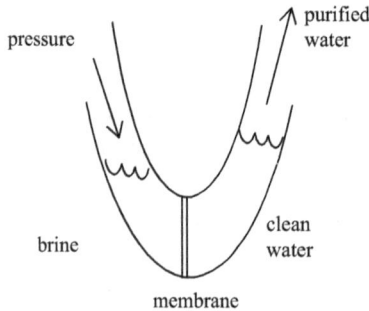

Figure 3.3: Reverse osmosis.

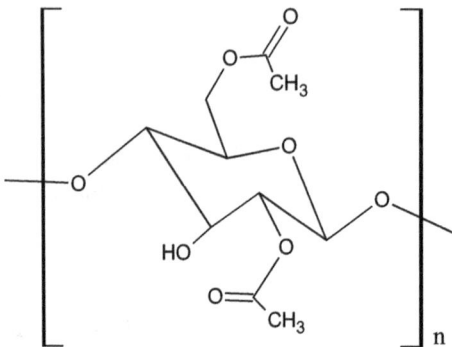

Figure 3.4: Lewis structure of cellulose acetate.

Dramatic examples of the results of reverse osmosis water purification have been highlighted in popular news programs. For instance, such demonstrations feature water that has been reclaimed from toilet flushes and is served to people on the street, all while being recorded. The idea behind such popular mentions of reclaimed water is to prove that the reclamation process makes the water clean, bacteria-free, and pleasant to consume, and is aimed at overcoming people's normal aversion to water that has had human waste in it at some point.

Estimates vary, but in general, it is accepted that cleaning and reclaiming water in this fashion is roughly an order of magnitude more expensive than simply minimizing leaks in a large distribution system.

Third: Desalination

Saline water can be desalinated in several ways, including flash distillation and the just-mentioned reverse osmosis, both of which result in clean water but also produce a more saline, brackish fraction of the original sample, which is usually returned to a greater water source, such as the ocean. Figure 3.5 shows the steps in such a process but places the highly brackish final fraction as a co-product. This will have to be the case in any such separation in a lunar environment.

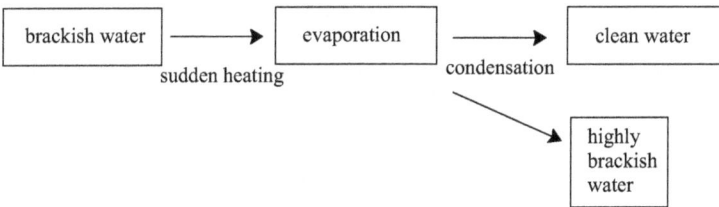

Figure 3.5: Desalination steps.

Once again, cost estimates vary when discussing desalination because such water purification systems are coupled to different types of distribution systems. For example, the desalinating plants in Singapore and those in Venezuela are integrated into distribution systems with different lengths from source to tap, and variations in operation. But in general, the cost here is another order of magnitude higher than the other means we have just discussed.

Since water on the lunar surface has been detected, but its salinity has not yet been studied in detail [11, 12, 14], it is unknown how brackish or saline such water is. When it is known how pure such water is, a more detailed plan of action can be formed regarding what is needed to make such water potable and drinkable.

Since the creation of a more brackish fraction than what was originally present is not necessarily a feasible option, it can be speculated that any lunar water may be flash-distilled, and the brackish fraction used for some other purpose. This form of gray water may find use in some more industrial application or may conceivably be used for plant growth, assuming those plants grown are strains and cultivars that have been found to grow in relatively saline waters in various environments on Earth [24–26].

References

[1]   Fluxes of fast and epithermal neutrons from lunar prospector: Evidence for water ice at the lunar poles, *Science*. https://www.science.org/doi/10.1126/science.281.5382.1496.

[2]   H. Lin, et al. In situ detection of water on the Moon by the Chang'E-5 lander, *Science*, Website. Accessed 10 April 2025, as: https://www.science.org/doi/10.1126/sciadv.abl9174.

[3]   Heterogenous distribution of water in the Moon, *Nature Geoscience*. https://www.higp.hawaii.edu/~gjtaylor/GG-673/Moon/Robinson+Taylor(2014).pdf.

[4]   H. He, et al. A solar wind-derived water reservoir on the Moon hosted by impact glass beads. Website. (Accessed 10 April 2025, as: https://www.nature.com/articles/s41561-023-01159-6).

[5]   A. Gasparini, and L. Wasser Water & ices on the Moon. – NASA science. Website. (Accessed 10 April 2025, as: https://science.nasa.gov/moon/moon-water-and-ices).

[6]   M. Anand, R. Tartese, and J.J. Barnes. Understanding the origin and evolution of water in the Moon through lunar sample studies, *Philosophical Transactions A: Mathematical Engineering Sciences*, 2014, Sep 13;372(2024):20130254. Website. Accessed 9 April 2025, as: https://www.ncbi.nim.nih.gov.

[7]   NASA. F. Chou. NASA's SOFIA discovers water on sunlit surface of Moon. Website. (Accessed 10 April 2025, as: https://www.nasa.gov/news-release/nasas-sofia-discovers-water-on-the-sunlit-surface-of-the-moon/#).

[8]   European Space Agency. ESA – Hunting out water on the Moon. Website. (Accessed 10 April 2025, as: www.esa.int/ESA_Multimedia/Images/2020/03/Hunting_out_water_on_the_Moon).

[9]   ESA – Water and oxygen made on the Moon. Website. (Accessed 10 April 2025, as: https://www.esa.int/Abour_Us/Business_with_ESA/Business_Opportunityes/Water_and_oxygen_made_on_the_Moon).

[10] NASA Science. How Ingredients for Water Could Be Made on the Surface . . . Website. (Accessed 10 April 2025, as: https://science.nasa.gov/solar-system/moon/how-ingredients-for-water-could-be-made-on-the-surface-of-the-moon).

[11] NASA Science – There's Water on the Moon? Website. (Accessed 10 April 2025, as: https://science.nasa.gov/solarsystem/moon/there's-water-on-the-moon).

[12] Space.com. There's lots of water on the moon for astronauts. But is it safe to drink? Website. (Accessed 10 April 2025, as: https://www.space.com/moon-water-astronauts-aqualunar-drinking-safety-contest).

[13] The Planetary Society. J. Mehta. Your guide to water on the Moon. Website. (Accessed 10 April 2025, as: https://www.planetary.org/articles/water-on-the-moon-guide#).

[14] S. Li, and R.E. Milliken. Water on the surface of the Moon as seen by the Moon Mineralogy Mapper: Distribution, abundance, and origins, *Science*, Website. Accessed 10 April 2025, as: https://science.org/doi/10.1126/sciadv.1701471.

[15] Astronomy Magazine. The Moon has less water than we thought. Website. (Accessed 10 April 2025, as: https://www.astronomy.com/science/the-moon-has-less-water-than-we-thought/).

[16] Tucson Water Department. Website. (Accessed 10 April 2025, as: https://www.tucsonaz.gov/Departments/Water).

[17] Philadelphia Water Department. Website. (Accessed 10 April 2025, as: https://water.phila.gov/pool/files/home-water-use-ig5.pdf).

[18] American Water Works Association. Website. (Accessed 10 April 2025, as: https://www.awwa.org/).

[19] National Association of Clean Water Agencies. Website. (Accessed 10 April 2025, as: https://www.nacwa.org).

[20] Water Quality Association. Website. (Accessed 10 April 2025, as: https://wqa.org/).

[21] European Water Association. Website. (Accessed 10 April 2025, as: https://www.ewa-online.eu/).

[22] British Water. Website. (Accessed 10 April 2025, as: https://www.britishwater.co.uk/).

[23] Australian Water Association. Website. (Accessed 10 April 2025, as: https://www.awa.asn.au/).

[24] D.G. Nocera. Artificial Leaf. Website. (Accessed 10 April 2025, as: https://chemistry.harvard.edu/people/daniel-g-nocera#:~:text=Nocrea%20has%20accomplised%20the%20solar.efficiencies%20of%20greater%20than%2010%25).

[25] A solar wind-derived water reservoir on the Moon hosted by impact glass beads, *Nature Geoscience*, 2023;16:294–300. https://www.nature.com/articles/s41561-023-01159-6.

[26] Wastewater treatment and water reclamation, A.J. Englande Jr. Published online 2015 Jul 22. doi: 10.1016/B978-0-12-409548-9.09508-7.

Chapter 4
Food

4.1 How can shelf-stable food be imported from Earth?

It seems painfully obvious, but at the outset of establishing any long-term human presence on the Moon, all food will have to be brought from Earth. This then becomes a matter of what foods are shelf-stable for the longest periods of time, and what preservatives and processes make them so. Table 4.1 provides a non-exhaustive listing of foods that are known to be stable for long periods of time and that should remain stable when transported from Earth. Existing organizations that utilize food packaged for long-term storage and future use include the militaries of many nations and rescue organizations, since they all require food for their personnel who may be deployed far from home or for the victims of some disaster. Routinely, all of these needs are for people far from where locally grown and farmed foods can be obtained. Additionally, NASA and other space agencies have examined the production of stable foods that can be stored at the space station and any other ships that we put into orbit [1–9].

Table 4.1: Longest shelf-stable foods.

Food	Shelf-stable time (yr)	Preservatives	Comments
Beans, dried	>30	None	
Ramen noodles	10	None	Routinely packed dry
Dry pasta	6–8		Can be indefinite, if dry
Tuna, canned	5		
Dried oatmeal	½		
Olive oil	1½–2	Keep in dark storage	
Peanut butter	3–5	Emulsifier	
Salt	Indefinite	N/A	
Honey	1 K	None	Edible honey has been found in 3,000-year-old tombs
Cheeses, e.g., cheddar, parmesan, gouda	½ to 1	Salt	Can be waxed for longer storage

It is noteworthy that peanut butter and canned tuna are listed in the above table and, indeed, are found on virtually all lists of highly shelf-stable foods, since these two are

https://doi.org/10.1515/9783111387307-004

important sources of protein. Various types of beans are also good protein sources, and have extremely long shelf-lives. The subject of protein in diets becomes extremely important for individuals living in an environment in which the gravity is only one-sixth that of Earth, and a continuing concern for the loss of muscle mass over time is present.

Additionally, foods like tuna and cheese contain significant amounts of calcium, which is also important, as the loss of bone density due to long periods of living on the Moon remains a studied subject, though one of some debate [7–10]. It is generally believed that osteoporosis cannot be reversed, but it can be slowed. What are now considered normal procedures involve lifestyle changes, possibly medication, and adjustments in diet. Besides calcium ingestion, adequate doses of vitamin D become important because vitamin D helps in the absorption of calcium. The idea of adjustments in lifestyle routinely means increases in anaerobic exercises – weight training – because many people do not normally perform any. However, any personnel living on the Moon will most likely be in excellent physical condition already and probably will have incorporated weight training of some sort into their daily routine.

Additionally, in examining all food types, carbohydrate sources such as honey are extremely important for energy production by personnel living on the lunar surface (refined sugar can also be stored for long periods, although not as long as honey). Carbohydrate breakdown, from glucose to pyruvate, is a main source of short-term energy for any person, with the breakdown of lipids (fats) being the human body's energy source for the longer term. Glucose is routinely converted to two moles of ATP in the process. Figure 4.1 shows this breakdown as a reaction.

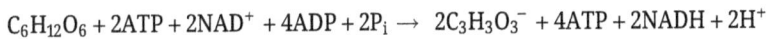

$$C_6H_{12}O_6 + 2ATP + 2NAD^+ + 4ADP + 2P_i \rightarrow 2C_3H_3O_3^- + 4ATP + 2NADH + 2H^+$$

Figure 4.1: Carbohydrate breakdown.

The breakdown of proteins for energy is not routine for human beings unless the person has begun starving. Usually, protein consumption is intended so that amino acids can be rearranged, and nitrogen-containing molecules can be used in muscle production and the uptake of elemental nitrogen.

Examples of foods that are shelf-stable for relatively long periods of time, routinely years, have been listed in Table 4.1 but can already be found in the variety of military rations throughout the world. These tend to be high-calorie meals, simply because the needs of soldiers in the field are such that these types of meals are necessary while they perform their missions. Table 4.2 shows a breakdown of the current U.S. military "meals, ready to eat." Table 4.3 shows a similar breakdown for the German Army, the Bundeswehr [11, 12].

Perhaps obviously, the palette of different people living and working in a lunar environment will have a certain amount of cultural bias, and may thus influence what meals are brought up from Earth.

Table 4.2: U.S. military field rations.

Meal	Components	Comments
Breakfasts	Egg omelets	High in protein
	Ham/pork sausage	
	Creamed ground beef and biscuits	
	Dried fruit bars	
	Cereals	
Lunches	Hamburgers	Balance of protein
	Chicken breast	and carbohydrates
	Spaghetti	
	Turkey slices	
	American or Swiss cheese	
	Bacon and cheese	
	Rice	
Dinners	Canned beef or pork	Balance of protein, fats,
	Biscuits	and carbohydrates;
	Chocolate bars	good
	Bouillon cubes	for energy
	Crackers	
	Peanut butter and jelly	

Table 4.3: Bundeswehr field rations (Einmannpackung), one day.

Category	Entré
Main item	Goulash
	Soups
	Noodles
	Tinned rye bread
	Sausages
Snacks	Chocolate
	Energy bars
	Sweet crackers
	Dried fruit mix
Drinks	Coffee powders
	Tea bags

$$C_6H_{12}O_{6(s)} + 6O_{2(g)} \longrightarrow 6CO_{2(g)} + 6H_2O_{(l)} + \text{energy}$$

Figure 4.2: Carbohydrate breakdown

Figure 4.2 illustrates the routine manner in which a generic reaction for the breakdown of carbohydrates is depicted. This is essentially represented as the reverse of photosynthesis. Sugars, with glucose being the most common, are consumed in the formation of energy in the form of ATP (as seen in Figure 4.1), which powers a variety of cellular chemical reactions. However, in this environment, it is also noteworthy that the CO_2 produced will be valuable and will be captured for some future use.

In looking at other environments in which CO_2 concentrations are important and are monitored, we can mention submarines, and the CO_2 produced by their crews. On a submarine, this gas is considered a waste product, and has to be scrubbed from the atmosphere, to prevent illness among the crew. On a lunar base, this will have to be "scrubbed," but in a different manner than on a submarine. Any lunar base will need to concentrate CO_2 for later use in any greenhouses, for plant growth. What was a waste product, released into the waters of the oceans, will become part of a loop, in this case, the part that makes photosynthesis possible.

Cost of lifting foodstuffs to a Moon base

The cost of lifting any material to a lunar colony will always be an important factor in determining what is chosen – and will always be part of the supply loop from Earth to the Moon. Therefore, the mass of a single meal can serve as a basis for how much mass will have to be lifted when we consider the number of people to be fed, the amount of time they will be off Earth, and thus how many total meals will be required. Assembling this into a formula, we can arrive at:

$$(\text{mass of meal}) \times (\#\ of\ \text{personnel}) \times (3\ \text{daily meals}) \times (\text{days on the Moon})$$

$$= \text{total mass}$$

It will, hopefully, be obvious that the number of personnel in any lunar base becomes the lynch-pin of the equation. As well, it is prudent to overestimate the total mass by some percentage, since personnel will be hungry after particularly strenuous work shifts, and may wish to have greater than the amounts afforded in pre-packaged, pre-determined meals.

Food dehydration

Foods imported to the Moon will, in many cases, be dehydrated, especially if water already on the Moon can be purified to the point where it is fit for human consumption. Thus, to an unknown extent, the consumption of dehydrated foods will depend upon the discovery and purification of sufficient water on or near the lunar surface. Put simply, the more water that is found and used on the Moon, the more any food being imported can be dehydrated or freeze-dried.

Breaking down food dehydration into steps, we see that existing food dehydrators work as follows:

1. Food is sliced, generally into even pieces.
2. The pieces are placed on trays in an air stream.
3. The airstream is maintained between 40 °C and 70 °C using one or more heating elements, and the stream is kept blowing for hours to days, depending on the thickness of the food being dehydrated.
4. Removing water from the food ensures that yeasts and molds do not grow on it.
5. Often, the food is immediately vacuum-packed to further ensure that no airborne organisms come into contact with it.

In this process, it may seem obvious, but even if a certain amount of fresh food can be grown in a lunar colony, a significant amount of dehydrated food will still have to be lifted to such a colony or habitat. Numerous companies already exist which market a wide variety of dehydrated foods to the public, for campers, outdoorsmen, and survivalists. Additionally, contractors to the United States Department of Defense are already capable of large-scale dehydration for the production of what are called "Meals, Ready to Eat." Other such companies provide combat rations to the militaries of Britain, France, Germany, and China, to name only a few of the nations with highly developed militaries.

4.2 How will food be grown on the Moon?

While large amounts of food will have to be brought from Earth to the Moon, the idea of growing food on the Moon, at least in part, solves the problem of the cost of lifting it. Unfortunately, local growth of a variety of plants also presents its own challenges. Plants that are currently grown hydroponically on Earth in the greatest amounts include those listed in Table 4.4, but this is not an exhaustive list.

Table 4.4: Hydroponic plant growth [13].

Plant type	Time to maturity (weeks)	Comments
Basil		Small space requirements
Bell peppers	7–12	Grown year-round
Cucumbers	7–10	Require more space than others in this table
Lettuce	6–8	Multiple annual harvests
Radishes	3–5	Some varieties in as little as 2½ weeks
Spinach		High in nutrients
Strawberries	12–16	
Tomatoes	8–12	

4.2.1 Most production of oxygen

It may not seem immediately obvious, but the food crops that can be grown in any lunar environment or habitat will be doubly useful if they also produce significant amounts of oxygen. Any plants grown should produce food, thus lowering the burden of lifting food from Earth to the Moon, and providing the added benefit of having fresh food for consumption by anyone living in a lunar base or colony. The reverse of Figure 4.1 is the simplified production of plant material from carbon dioxide and water, as shown below in Figure 4.3. Depending on a number of factors, this can be starch – edible and digestible by humans – or cellulose, which is not edible but does represent a sink for carbon dioxide that would otherwise need to be removed from the atmosphere of any lunar dwelling spaces.

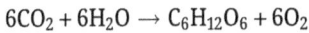

$$6CO_2 + 6H_2O \rightarrow C_6H_{12}O_6 + 6O_2$$

Figure 4.3: Production of starch or cellulose.

Of interest to us here is not only the production of food but also the co-production of oxygen. We have already discussed the obvious, vital need for oxygen in Chapter 2 but reiterate it here, as plant growth will alleviate some of the need for its import from Earth.

4.2.2 Hydroponic gardening

One means of producing edible plants in any lunar colony is through hydroponic gardening. This involves growing plants in aqueous solutions that contain nutrients but do not involve soil. It is currently unknown whether the lunar regolith contains nutrients that will aid in plant growth. If it does not, the introduction of hydroponic farming may become a viable option. Hydroponic farming is well-established enough that organizations exist to advocate for it [13–16].

Currently, hydroponic farming is relatively expensive but is sometimes utilized in the gourmet dining market, where the meal brought to the table must be attractive, without any blemishes caused by prolonged contact with soil or by insects. The cost of hydroponic farming may be outweighed by the savings involved in having produce grown right where individuals living in a lunar colony reside.

4.2.3 Greenhouses

Greenhouses are used extensively in places such as the Netherlands and Iceland, both being nations with relatively short growing seasons. Greenhouses also exist in numerous other places on Earth, not all of which are areas with short growing seasons or scant rain. Designs for greenhouses can vary widely, but a basic schematic, as shown in Figure 4.4, illustrates the fundamental components that are common to virtually all of them, including ample glass surfaces to allow sunlight in, trays or beds for plant growth, and a water source to keep plants growing on a schedule that is as normal as possible when compared to their growth outdoors.

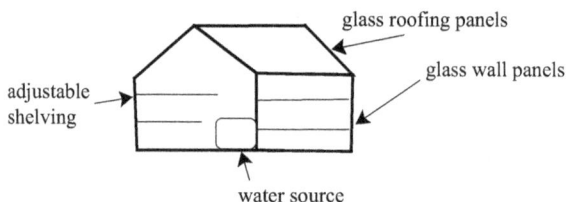

Figure 4.4: Greenhouse schematic.

On purpose, our schematic looks roughly like many greenhouses that already exist, with a roof of glass paneling and straight walls, also of glass paneling. In a lunar habitat, any greenhouse that can be constructed may simply have double-thick glass in any spot – most likely tempered glass or laminated glass – and may have concave or convex glass paneling, so that the difference between the relatively high pressure of the greenhouse and the vacuum surrounding it can be accounted for. Any lunar greenhouse operation may not look like the familiar structures on Earth but will need to be well-adapted to prevent heat transfer and to allow maximum light penetration for plant growth.

4.2.4 Vertical farming

The idea of vertical farming is one that has several decades of history, but it still has the potential to expand widely in places on Earth where land is at a premium or where arable land is scarce [15, 17–20]. The simplest example of vertical farming may be the one enacted by do-it-yourselfers, who pile up used tires, fill the central hole with soil, and plant potatoes at each layer as the stack of tires is created. Figure 4.5 shows a schematic.

In a vertical farm of this size and simplicity, plants such as potatoes can be planted while the stack is being constructed and harvested as the stack is taken apart. This certainly saves space, and tends to save water as well, although water use cannot be totally eliminated. Both will be at a premium in a lunar habitat. Additionally, such plant growth will alleviate the problem of carbon dioxide build-up from other activities, since it is essential to plant growth.

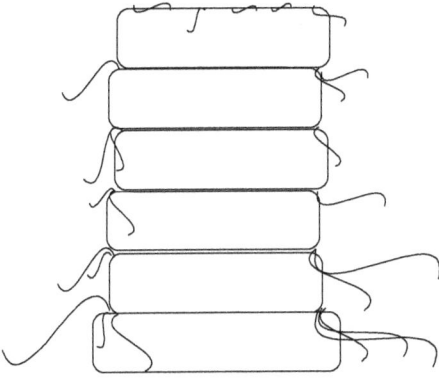

Figure 4.5: Vertical farming tire stack.

But vertical farming has become much larger than home projects and has expanded into associations that advocate for it [15, 17–20]. For example, in the past five years, the largest vertical farm in the world has been constructed in Dubai, simply because fresh water and arable land are scarce there, and the need for vegetables, perhaps obviously, rises as the number of people living in an area increases – Dubai being a nation with a very high population density (circa 750 people per square kilometer). This type of vertical farming uses a skyscraper as more than a building in which residences or businesses are housed. The skyscraper becomes a means by which edible plants are grown in a very limited space. The study of examples of successful vertical farming projects on Earth will be of great help should this farming technique be developed for a lunar habitat.

4.2.5 Storage of foods grown on the Moon

We can theorize that any plants grown in a lunar colony will not need to be stored for long periods of time. Almost all can be grown to be eaten, and thus some short-term refrigeration might be the only requirement for any of it to be stored. It is more conceivable that such produce will be eaten within days or hours of being harvested, whether from hydroponic bays, greenhouses, or vertical farms, thus limiting the need for any long-term storage or refrigeration system.

References

[1]   Defense Logistics Agency, Operational Rations Website. (Accessed 10 April 2025, as: https://dia.mil).
[2]   Armed Forces Pest Management Board, Technical Guide No.38 – Protecting Meal, Ready-to-Eat Rations and Other Subsistence During Storage. Website. (Accessed 10 April 2025, as: https://www.acq.osd.mil).

[3] How Stuff Works. How MREs Work. Website. (Accessed 10 April 2025, as: https://science.howstuff works.com/mre3.htm).

[4] NASA: Space Food and Nutrition. Website. (Accessed 10 April 2025, as: https://www.nasa.gov/wp-content/uploads/2009/07/143163main_space.food_.and_.nutrition.pdf).

[5] BBC. What will we eat on the Moon? The food is literally out of this world. Website. (Accessed 10 April 2025, as: https://www.bbc.com/travel/article/20240525-what-will-we-eat-on-the-moon-the-food-is-literally-out-of-this-world).

[6] NASA. Space Food Systems. Website. (Accessed 10 April 2025, as: https://www.nasa.gov/director ates/esdmd/hhp/space-food-systems/#:~:text=As%20an%20example%2C%20beverages%20start, products'%20acceptability%20and%20shelf%20life).

[7] The European Space Agency, Food. Website. (Accessed 10 April 2025, as: https://www.esa.int/En abling_Support/Preparing_for_the_Future/Space_for_Earth/Space_for_health/Food).

[8] The European Space Agency: From tubes and cubes to haute cuisine – the refinement of space food. Website. (Accessed 10 April 2025, as: https://www.esa.int/Science_Exploration/Human_and_Ro botic_Exploration/Business/From_tubes_and_cubes_to_haute_cuisine_-_the_refinement_of_ space_food).

[9] CAS. A Division of the American Chemical Society. Space food for artemis has real-world applications. Website. (Accessed 10 April 2025, as: https://www.cas.org/resources/cas-insights /space-food-artemis#:~:text=NASA's%20Artemis%20Program%20is%20an,nutrition%20in%20Earth's %20challenging%20environments).

[10] Counteracting bone and muscle loss in microgravity. Website. (Accessed 10 April 2025, as: https://www.nasa.gov/missions/station/iss-research/counteracting-bone-and-muscle-loss-in-microgravity/).

[11] Defense Logistics Agency. Website. (Accessed 10 April 2025, as: https://www.dla.mil/Troop-Support /Subsistence/Operational-rations/MRE/).

[12] Bundeswehr Subsistence Office. Website. (Accessed 10 April 2025, as: https://www.bundeswehr.de/ en/organization/infrastructure-environmental-protection-and-services/bundeswehr-subsistence-office).

[13] Hydrobuilder. Website. (Accessed 10 April 2025, as: mygardyn.com – and – worldgraceproject.org – and – https://hydrobuilder.com).

[14] Hydroponic Society of America. Website. (Accessed 10 April 2025, as: http://www.hydroponicsocietyo famerica.org/).

[15] Association for Vertical Farming. Website. (Accessed 10 April 2025, as: https://vertical-farming.net/).

[16] Organic Farmers Association. Website. (Accessed 10 April 2025, as: https://organicfarmersassociation.org).

[17] Just Vertical, Inc. Hydroponic Farming Experts. Website. (Accessed 10 April 2025, as: https://commercial.justvertical.com).

[18] Maine Organic Farmers and Gardeners. Promoting Organic Agriculture Through Education, Training and Advocacy, Hydroponic. Website. (Accessed 10 April 2025, as: https://mofga.org).

[19] Association for Vertical Framing. Website. (Accessed 10 April 2025, as: https://vertical-farming.net).

[20] USVFA, U.S. Vertical Farming Association. Website. (Accessed 10 April 2025, as: https://www.usvfa.org).

Chapter 5
Living space

It is incredibly obvious to state that, when it comes to living space, every option for some form of human habitation on the Moon involves material that will have to be brought from Earth. Some form of planning for living space on the Moon now goes back decades [1–5], and generally falls into a few categories. Dividing into two major ones, we will have to live on the surface or under it. Dividing this further, surface habitation will have to be made from rigid materials or what can be called inflatable materials, either of which must be impermeable to gas migration. Sub-surface living will always involve some sort of cave structures but can be divided into caves with ceilings, walls, and floors that are lunar rock, or those that are some sort of inserts or liners which are made on Earth and transported to where they will be used, to the caves into which they will be inserted. Table 5.1 gives a broad look at the possibilities of lunar habitats.

Table 5.1: Lunar habitat possibilities.

Type	Details	Comments
Surface		
Surface, inflatable	Multi-layered wall materials	Collapsible, ease of transport to the Moon
Surface, rigid	Resistant to temperature and flying "debris"	May be made using regolith
Surface, stationary	Rigid materials, self-contained systems	For long-term presence in one spot
Below surface		
Caves	Can be bored or natural	Require collapsible/inflatable linings
Hybrid cave and surface		Combines details of all of the above

5.1 What insulation will be needed?

Existing models
It is almost a certainty that the living space in any initial lunar colony will best be described as crowded. When the Apollo missions landed on the Moon in the late 1960s and early 1970s, the lunar capsules that landed were not much more than cramped spaces in which to sit, and store necessary equipment. Eating and other bodily requirements were part of the astronauts' suits. And while there have been significant

https://doi.org/10.1515/9783111387307-005

advances in materials chemistry, engineering, and the ability to form a pressurized space, there are still significant challenges in creating spaces and areas large enough for people to live in for prolonged periods of time.

Because of the challenges in creating, defining, and building living spaces for any lunar presence, it is therefore logical to examine existing habitats on Earth that utilize space to a premium, take into account heat transfer between one side of a wall and another, or the lack thereof, and determine if there are common factors between them.

Submarines

Submarines are capable of withstanding extreme pressure from the outside pushing inward, as opposed to pressure pushing from the inside outward, as would be the case in any lunar habitat. It is, therefore, not a perfect model to examine submarines and their living environments and simply think of the challenges in reverse. Despite this, there are certain similarities that should be examined. Additionally, there is a significant amount of information on how submarines function, since they have been used in both war and peace for over a century [6, 7].

Submarines are double-hulled because the space between the hulls is flooded during submersion to make them function under the surface of the oceans (no engine is strong enough to simply drive and force a submarine under the surface of the ocean). The reverse of diving – when a submarine surfaces – requires the water between the hulls to be blown out and thus requires air pressure to drive the water out. In the case of a lunar module that will be used for a prolonged time, the idea of a double "hull" or wall has some merit, since this can function as an adiabatic wall.

Almost all submarines are constructed with hulls made of steel, although the Soviet Union did have an alfa-class series of submarines which were constructed with titanium hulls. The reason for the titanium design was that such submarines were supposedly able to dive deeper and run swifter than other, steel-hulled submarines. It can be assumed that beyond this, such submarines functioned as did all others, although some material related to their construction may still be classified. Yet, similarities between submarines and a lunar habitat should include the following:

First, for any lunar dwelling or workspace, should there be any kind of leak or malfunction in one wall, the idea of a second wall to keep the atmosphere in place will prevent an immediate, catastrophic loss of atmosphere. The time saved and limitation of the problem can then be used to repair the rip or puncture.

Second, much like submarines, it will be safer to have the double walls extend for only small portions of the exterior of a habitat, into compartments. Compartmentalization will also work toward keeping any loss of atmosphere to a minimum.

Third, a vacuum between the two walls or hulls is useful in preventing heat transfer from any habitable area to the lunar exterior. Utilizing a vacuum in this manner can also be considered cost-effective, since, in such a case, no gas will have to be imported to serve as insulation for this "in-between" space. However, to ensure a continuous vacuum, some form of pump or pumping system will most likely be required, pumps that will have to be imported from Earth.

When the double-wall design functions properly – especially with a vacuum in be-tween – it is an adiabatic wall, preventing the transfer of heat from any habitat mod-ule to the lunar surface. As mentioned, pumping apparatus will still have to be em-ployed and maintained to prevent any small leaks that may occur from becoming a larger problem.

Fourth, if a gas between the inner and outer walls is to be used, this gas in the space between the two walls can serve as an insulator. This takes away some of the burden on a heating apparatus in any habitable area.

Finally, the double-wall design could serve as a storage space for any material that can withstand the cold and vacuum of space, since it could rest against the outer wall. Figure 5.1 shows a bow-to-stern simplified schematic.

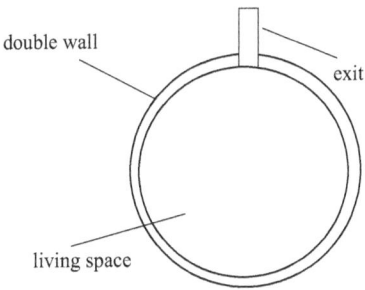

Figure 5.1: Submarine double-hull.

Airplanes

In a similar manner to submarines, but perhaps more akin to a living space surrounded by a vacuum, passenger airplanes function by maintaining a pressurized cabin when the environment outside the plane has a much lower atmospheric pressure. Figure 5.2 shows a schematic of this, indicating that there are two outer layers to the plane and illustrating the seats and storage space within. When riding in an airplane, most passen-gers do not consider the fact that both the walls of the plane, as well as the windows, are two layers thick. A close examination of any airplane window will reveal that there are two layers of glass present. This design prevents heat transfer and provides some measure of protection should the outer layer crack or break in an accident.

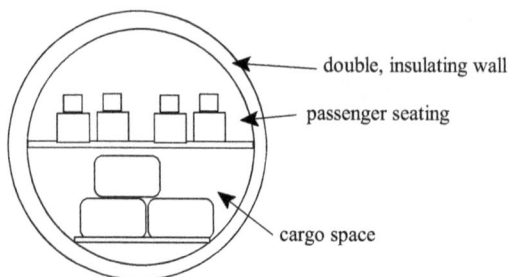

Figure 5.2: Airplane cross-section.

Antarctic research stations

The punishing cold of the Antarctic, and the need for people to be present there year-round, provide a third possibility for an analogous situation when examining the construction of living and working spaces on any lunar station. Examples include virtually all of the stations manned on the southern-most continent. We will focus on only a few.

The Halley VI research station [8]

The Halley Research Station, administered by the British Antarctic Survey, has features that may be very useful in any lunar dwelling. Importantly, it is a series of modules that *are all on hydraulic legs*, so that it can be moved if necessary. This may prove to be an important aspect of a lunar base, since discoveries worth living near may not all be in the same area, and new discoveries may require moving personnel, as well as living and working spaces.

While much of the structural material of the Halley Research Station is made of steel, glass-reinforced plastic is used as well to insulate the station's modules from the cold. This principle will be the same as that for any outer wall of any lunar habitat or for any area where windows are required. Figure 5.3 shows the basic design of one of the Halley VI modules. It is noteworthy that the shape of the habitats of Halley VI is designed to withstand the fierce Antarctic winds. This will not be a concern on the lunar surface, but the impact of micro-meteors will have to be studied.

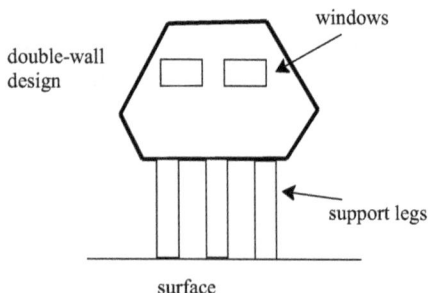

Figure 5.3: Antarctic living module.

The Australian Antarctic Building System (AANBUS)

Australia is a nation that has had a long presence in Antarctica and has been building there since the 1950s [9]. Current stations have utilized steel supports on concrete bases – which may be possible for a lunar habitat – but layered panels as outer wall material. Australia has addressed the problem of using concrete, a material that requires water for its setting, in an environment so cold that water freezes before the concrete can set. This may not translate directly to construction in a lunar base but can still provide some useful insights.

In the case of the Australian Antarctic structures, insulating panels that have been installed can be removed for maintenance, which may also prove adaptable to a lunar environment. Once again, the possible impact of micro-meteors dictates that walls or panels that make up the outer layer of a habitat be either removable as a type of maintenance or made of some material that can be patched on the spot.

5.2 What lightweight materials are optimal?

Moldable plastic

A variety of plastics exist on Earth, which have been scaled up in production to tens of millions of tons. Polyethylene (PE) is the largest single plastic in terms of production, with over 100 million tons currently made per annum. Figure 5.4 shows the repeat unit of PE, including one example of a straight chain and one example inclusive of some branching, since it is the branching that defines the differences between low-density (LDPE) and high-density (HDPE) varieties of this plastic.

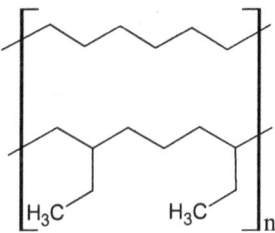

Figure 5.4: Polyethylene structure.

If used in any building applications for lunar habitats, PE will simply have to be imported from Earth. Pure PE is actually rather porous to some gases, and thus high-density PE or cross-linked PE will be more suitable in lunar construction.

The generally low density of PE makes it a good candidate for many applications because the cost of lifting it will be lower than that of other materials. Additionally, the strength of Tyvek – a commercial form of HDPE made from spun fibers – is such that layers of it may prove very useful in making walls and other support structures for any living space.

For complete impermeability to gas movement, HDPE has, in the past, been layered with other plastics. Ethylene vinyl alcohol co-polymer (EVOH) has been used successfully in the past in this respect.

Aluminum/magnesium

Both aluminum and magnesium, as well as aluminum-magnesium alloys, find use in the aircraft industry simply because they are low-density metals and can form strong alloys. Even if lodes of ores such as bauxite are found in abundance on the lunar surface, however, it will most likely be economically more advantageous to import aluminum items as finished parts from Earth. This is because of the high energy requirements needed to extract aluminum metal from the ore. Figure 5.5 shows the basic chemistry for bauxite refining as two half-reactions, known as the Hall-Héroult Process.

$$Al^{3+} + 3e^- \rightarrow Al_{(l)}$$

$$2\,O^{2-} + C_{(s)} \rightarrow CO_{2(g)} + 4e^-$$

Figure 5.5: Hall-Héroult process.

Note that the Hall-Héroult Process utilizes carbon anodes as a reducing agent. These are consumed in this high-energy reaction and must be replaced as the process continues. There is currently no reported reduced carbon on the lunar surface in any large amount from which to make such anodes.

5.3 Are underground dwellings a living space possibility?

It seems logical that, if living on the lunar surface for any length of time poses dangers from solar radiation, it might be wise to construct underground living quarters. This is dependent upon what a specific layer of regolith is made of and whether it is possible to tunnel into the material.

A cave has been located in the Sea of Tranquility that is believed to be an ancient lava tube. Since lava tubes are formed by volcanic activity, it is logical to believe that others may exist somewhere on the lunar surface. NASA's Lunar Reconnaissance Orbiter (LRO) has been able to verify this cave's existence. The LRO was launched in 2009, and continues to be in lunar orbit today. It was not designed simply to look for caves, but rather to map the surface, with the goal being to locate areas that may be suitable landing sites (the subject of the next chapter).

5.3.1 Air passages versus compartmental units

In numerous science fiction movies, there are some forms of air passages in any alien habitat or spaceship. While this is convenient for moving a movie plot forward, it is not particularly realistic, since a leak anywhere in a connected space could affect the entire area. It is much more realistic to envision compartmental units, much like those on a submarine or those in an Antarctic station, with some type of air-tight door or portal between sections.

5.3.2 Wall and bulkhead materials

In living in some underground environment on the Moon, it is perhaps obvious that the walls, ceiling, and floors can simply be the regolith, as drilled or tunneled into. As long as they are tested to be air-tight, they should be sufficient for a living or working space. But it is also feasible to think of any such surfaces being lined with some form of plastic, at least as a secondary safety device, to further ensure that air does not escape. Plastics such as the already mentioned PE are relatively low in density, which is important since they will have to be imported from Earth. Additionally, such "plastic walls" can serve as a layer of insulation between any habitable area and the lunar rock. Virtually any type of plastic or composite of multiple plastics can serve this purpose, as long as it prevents the transfer of gas. When considering transport of them to the Moon, their densities become a factor to consider. Table 5.2 shows a non-exhaustive list.

Table 5.2: Plastics' densities.

Name	Abbreviation	Density (g/cc)
Polyetheretherketone	PEEK	1.30–1.32
High density polyethylene	HDPE	0.94–0.965
Polyvinylidene chloride	PVDC	1.63–1.80
Polyethylene terephthalate	PETE	1.38
Ethylene vinyl alcohol	EVOH	1.10–1.20
Polychlorotrifluoroethylene	Kel-F81	2.10
Poly-tetrafluoroethylene	PTFE	2.20

5.4 From what will surface stations be made?

Whether or not people in any lunar colony spend most of their time underground, there will have to be surface stations where observation and interaction with lunar activities and solar observations will occur. The wall materials we have just examined will definitely be used at such stations, but a great deal of glass chemistry will be important, since windows will look at our Sun without the benefit of any atmosphere and must be able to withstand the impact of tiny, fast-moving particles that strike the lunar surface.

A wide variety of materials can be incorporated into the production of glass. In what is commonly called "bulletproof glass," a layer of polycarbonate is sandwiched between two layers of traditional glass (SiO_2), with a thin layer of adhesive keeping the layers together. Figure 5.6 shows the basic scheme for this. To make such material even more resistant to impacts, multiple layers of this three-ply sandwich of glass and polycarbonate can be formed.

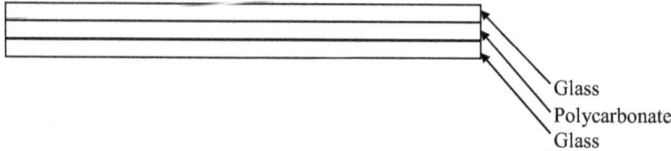

Glass
Polycarbonate
Glass

Figure 5.6: Diagram of bulletproof glass.

The idea of multiple layers of this highly resistant glass has a very practical application. Material such as micro-meteors impacts the Moon constantly, much as it does the Earth. However, since the Moon has no atmosphere to incinerate such particles, any glass window of a lunar habitat must be able to withstand direct impacts without sustaining serious damage. Table 5.3 lists several types of high-strength glass but is not inclusive of all possibilities.

Table 5.3: High-strength glass types.

Type	Comments
Ballistic glass	Usually incorporates layers of plastic and silica
Heat-strengthened glass	Made by slow cooling after intense heating
Laminated glass	A glass sandwich with a polyvinyl-butyral core
Polycarbonate glass	Can be used as a single plastic layer or as a sandwich
Tempered glass	Manufacture requires controlled heating, followed by slow cooling

Beyond the idea of glass that can survive multiple high impacts, it should be noted that any lunar habitat windows will probably need some form of "sun screen." In the past, what is called silvering has been applied to various types of glass as a thin layer. However, the Apollo astronauts' faceplates were made of Udel polysulfone (PSU), manufactured by Solvay, because it is both tough and able to withstand high temperatures. Figure 5.7 shows the repeat structure of PSU.

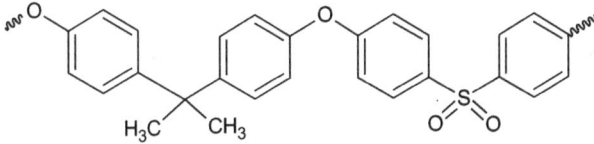

Figure 5.7: Polysulfone structure.

Since this is an already proven material in a previous space program, it is likely that it can be used in future, similar applications in any lunar habitat, although improvements have probably been made in the intervening decades. It may not yet be possible to form it in large sheets, or sheets large enough to make windows for any surface structure. However, in the 1960s, it was possible to make the faceplate for an astronaut's helmet. Extending this further should not prove to be an insurmountable challenge.

References

[1] G.B. Ganapathi, J. Ferrall, and P.K. Seshan. Lunar base habitat designs: Characterizing the environment, and selecting habitat designs for future trade-offs, *NASA JPL*, 1993. chrome-extension: //efaidnbmnnnibpcajpcglclefindmkaj/https://spacearchitect.org/pubs/NASA-CR-195687.pdf.

[2] F. Ruess, J. Schaenzlin, and H. Benaroya. Structural design of a lunar habitat, *Journal of Aerospace Engineering*, 2006, July:133–158.

[3] Internal Layout of a Lunar Surface Habitat. chrome-extension://efaidnbmnnnibpcajpcglclefindmkaj/ https://ntrs.nasa.gov/api/citations/20220013669/downloads/Internal%20Layout%20of%20a% 20Lunar%20Surface%20Habitat.pdf.

[4] H. Benaroya. Lunar habitats: A brief overview of issues and concepts, *Reach*, 2017, December;7– 8:14–33. https://www.sciencedirect.com/science/article/abs/pii/NASAS2352309318300014.

[5] L. Stahl. Company 3-D prints houses on Earth, partners with to 3D print infrastructure on the moon, *CBS News 60 Minutes*, 2023, 8 October.

[6] US Department of Defense, Living in the Deep. Website. (Accessed 10 April 2025, as: https://www. defense.gov/Multimedia/Experience/Living-in-the-Deep/#:~:text=America's%20submarines%20have %20come%20a,each%20other%20%E2%80%94%20and%20underway%20replenishments).

[7] Royal Navy. Submarine Service. Website. (Accessed 10 April 2025, as: https://www.royalnavy.mod. uk/careers/services/submarine-service).

[8] British Antarctic Survey. Halley VI Research Station. Website. (Accessed 10 April 2025, as: bas.ac.uk/ polar-operations/sites-and-facilities/facility/halley).

[9] AANBUS Buildings. Website. (Accessed 10 April 2025, as: Antarctica.gov.au/Antarctic- operations/stations/amenities-and-operations/buildings-and-structures/aanbus-buildings/).

Chapter 6
Landing ports

The idea of landing ports or specified areas for craft coming from Earth to the Moon, or for craft departing the Moon – and the need for such areas – may seem blatantly obvious. However, a significant number of challenges will have to be overcome to ensure that such zones are safe, easily reusable, and do not cause degradation or erosion to any structures, machinery, or equipment around them.

Again, perhaps obviously, we have already landed on the Moon several times during the Apollo missions. The first, Apollo 11, on July 20, 1969, was the most perilous simply because we did not know what the composition of the Moon's surface was in terms of the density and packing of material. Bluntly, it was unknown whether or not the Apollo landing craft would sink in to the lunar surface to a degree that it could not re-enter orbit, or if it would come to rest on the surface – as, indeed, it did.

But this example of how the Apollo space craft landed on the Moon is one of two broad means by which any craft can land on a surface. A craft can either drop down – as the Apollo craft did – or it can coast in, as aircraft do when they land on Earth. Each presents challenges, and those are different on the lunar surface than they are on Earth.

6.1 What will be the surface requirements for landing ports?

The following conditions will have to be met in developing any landing port on the lunar surface. At a minimum, NASA has determined that these include the following:
1. **Flat, level surface.** The size of the landing pad will depend on the size of the craft, as well as on whether such craft will approach vertically or at an angle.
2. **Fuel.** The fuel or propellant required both for a craft to land and for it to depart the lunar surface needs to be minimized, simply because the mass of an entire spacecraft, including the fuel, must be calculated for in any landing.
3. **Ease of a launch.** It is obvious that the craft must return from the surface to space. The ease with which any craft can break free of the limited but existent lunar gravity will be important when return trajectories to Earth are computed. This is also critical when a trajectory to a second craft, to which the first will dock, is computed.
4. **Presence of sunlight.** Any craft that lands or departs will most likely have to do so in an area of the Moon that is in direct sunlight. Lights can be part of any spacecraft, but natural sunlight is abundant and can thus help make any landing safer.

https://doi.org/10.1515/9783111387307-006

5. **Level regolith.** We have mentioned a flat, level surface, but starting with a level section of regolith eliminates the work that would otherwise need to be done to flatten an area.
6. **Communication links to Earth.** The site at which any craft will land or depart needs to be in direct communication with control centers on Earth.

One problem that may have to be addressed in some novel way is the clearance of any loose material from the landing area. On Earth, this can be done by blowing material off an airport runway or by vacuuming up material from such an area. Since both of these depend on a gas mixture – air, basically – some other method will have to be devised. This may be as direct as having robotic sweepers push loose debris out of the landing area. Whatever the case, though, what seems like a trivial problem is quite serious, since a landing spacecraft's exhaust always provides some amount of fast-moving gas that can send small particles racing with the speed and lethality of a bullet. Even if no personnel are nearby, this can cause significant damage to other machinery and equipment that is in the immediate vicinity [1–4].

Curiously, not only NASA, but also the European Space Agency (ESA) has put considerable thought into what is required of a landing port on any lunar surface. Summarizing this, we have the following:
1. Safety – once again, the area must be free of debris that could cause potential problems. ESA has suggested that landing craft be equipped with some form of LIDAR, to verify any suitable landing site.
2. Sunlight – this is not only because operations can be performed more safely in sunlight (perhaps obviously, artificial light can be provided for a dark-side landing, as mentioned), but because the temperature of the lunar night can reach −160 °C. Considering that the coldest austral winter days in Antarctica seldom go below a temperature of −80 °C, we currently have very little data on the performance of materials at temperatures this low.
3. Scientific merit – landing sites should be in close proximity to areas that are of high interest in terms of scientific study. Such areas can include the presence of water, a relatively high abundance of ^3_2He, or the presence of valuable minerals. In all cases, the presence of such materials should be proximate enough to the landing site or sites that samples can be retrieved easily – or that larger operations can commence.

Landing sites are, in some ways, connected to the type of craft that will be used. Much is being done to prepare the Artemis missions at this time, but the ESA is also working on the Argonaut craft, which should be able to land on a wide variety of surfaces, not all of them smooth and level. Since Argonaut is designed to carry up to 1,500 kg of personnel, equipment, or material – and to do so robotically if needed – its landing capability must be robust [2].

As well, Luna-27, the Russian spacecraft, is designed to drill into the regolith at the Moon's South Polar Region, primarily to determine the presence of water. This craft should be able to land on other-than-flat surfaces, especially since the south polar region of the Moon is considered to be a rough overall surface [5, 6].

It is easiest for any spacecraft to land on or take off from a naturally occurring flat surface, as opposed to one that has to be formed through human construction. Should such construction be necessary, two scenarios can be considered:

First, cutting, abrading, and smoothing a surface.
Examining this idea of abrading an area to make a flat surface, the cutting tools – basically, stone saws – will have to be imported from Earth, and be of a design that requires no liquid, combustible fuel to function. Electric motors charged by batteries will be required.

Cutting, abrading, and flattening a surface will leave debris behind. This must be cleared far enough from the landing area so that the exhaust from any craft does not blow it about. The lower gravity of the Moon, coupled with unanchored debris, could result in material being blasted about with the force of bullets or other projectiles, which can cause significant damage.

Second, the use of cement and concrete to form a flat surface.
Examining the idea of the pouring of cement, an obvious feature of virtually all cement poured on Earth is that it requires water. This will be a challenge in a lunar surface environment. It is imagined that, should the pouring of cement be required, the area into which it will be poured will first be enclosed so that any water does not immediately boil away into nearby space. The hardening of cement and concrete is an exothermic reaction, and any area in which this takes place will have to remain isolated with barrier walls and a roof until the material has set.

The chemical formula for cement components includes what is listed in Table 6.1. To create concrete, stone is often incorporated. On the lunar surface, this could conceivably include loose lunar regolith. What is not shown in Table 6.1 is water.

Table 6.1: Cement components.

Formula	Name	Abbreviation
$3CaO \cdot SiO_2$	Tricalcium silicate	C3S
$2CaO \cdot SiO_2$	Dicalcium silicate	C2S
$4CaO \cdot Al_2O_3 \cdot Fe_2O_3$	Tetracalcium aluminoferrite	C4AF
$3CaO \cdot Al_2O_3$	Tricalcium aluminate	C3A

On Earth, what are called rapid runway repair kits are part of the support systems of the militaries of several countries. These are used when airstrips are compromised or destroyed in war and are utilized to resurface an area so that war planes can land safely. However, as already mentioned, these require water for concrete and cement formulations to set.

6.2 How will dust and debris be handled?

The movement of some amount of lunar surface material will result from craft landing and departing, with the exhaust gases they produce and emit being the driving force for such movement. Thus, some form of maintenance of any landing ports will be required, precisely because there is no atmosphere on the Moon to move material about. It should be kept in mind that the current lunar regolith has built up over billions of years, with nothing that can be called lateral movement, once again, simply because there is no gaseous atmosphere to push particles from one place to another.

Blowing and plowing

On Earth, blowing or sweeping areas clear, such as airplane runways or helipads, is relatively common. However, this process is dependent on the atmosphere and atmospheric gases moving any solid debris on the affected surface. Clearing and cleaning a lunar landing port will require something equivalent to a street sweeper (sometimes called a dust truck or a mechanical sweeper truck), one that can gather up debris without relying on an atmosphere. This debris may simply be moved to a designated collection location. Alternatively, it may be gathered by such trucks and later repurposed as construction material, perhaps for producing additional structures, such as walls or bricks.

Use of water

We have mentioned several times in this chapter that the creation of cement requires the use of water; but in many cases, water is also used to clean street and sidewalk surfaces – surfaces much akin to any lunar landing site. At this point, it simply seems impossible to propose using water in any sort of cleaning or maintenance operations for lunar landing sites.

Waterless landing site maintenance

The solution to maintaining any lunar landing site may be two-fold, both ideas being simple ones. First, something as simple as brushes or shovels can be repeatedly passed over any site to rid it of loose debris. Second, the slow descent of spacecraft, including the use of retro-rockets at high altitude, may prevent the movement of any material on the surface of the regolith.

6.3 Will re-usable space "tanks" for moving products to and from the Moon be possible?

All early rockets to the Moon, as well as those to Earth orbits, were aimed solely at getting the payload – often three astronauts – to their destination and through their mission, without much concern for any kind of economy in the use of the ship and materials involved in creating it. To be rather blunt, since these were single-use craft, a great deal of waste was generated, with some components being left on the lunar surface and some being allowed to burn up upon re-entry into the Earth's atmosphere. This cannot continue as we move toward a more permanent, cyclic form of transportation to the Moon. All vehicles departing Earth for the Moon will need to be robust enough to be indefinitely re-usable.

Table 6.2 is a non-exhaustive list of several metals, alloys, and materials that are low in density but can be high in strength. This will be important in building any craft that must land repeatedly on the Moon and re-enter the Earth's atmosphere. The idea we have repeated here, that lunar regolith debris might impact any craft, means that it should be built of corrosion and impact-resistant materials and be high in strength.

In the past, chromium alloys, as well as titanium alloys, have proven useful in regard to making lightweight, strong materials for aircraft, as well as in some water-craft. Additionally, ceramic tiles are listed here because they were used as parts of the skin of the U.S. Space Shuttle fleet: Atlantis, Challenger, Columbia, Discovery, and Endeavor. The Soyuz spacecraft is predominantly made of aluminum but also has ceramic materials as its outermost skin.

Table 6.2: Density of materials [7–9].

Metal	Density (g/cc)	Comments
Aluminum	2.7	
Magnesium	1.738	Often alloyed with aluminum
Lithium	0.534	Al-Li alloys can be low density and high strength
Titanium	4.51	Used as alloys in aerospace applications because of their high strength and low corrosion
Chromium	7.19	Ni-Cr and Co-Cr alloys are used in aerospace applications
Inconel 718		High strength Ni-Cr alloy
Directionally-solidified superalloys (DS)		Aligned grain structure for enhanced strength. Used in jet engine turbine blades
Ceramics	0.144–0.352	Termed high-temperature reusable surface insulation (HRSI) for the Space Shuttles

The U.S. space shuttles, launched by NASA, the first in 1981, certainly had some of the properties shown in Table 6.2 when they were constructed. Small particles exist in near-Earth orbit, which can be moving fast enough to damage any ship not made of high-strength metals. However, the shuttles, though made for repeat use, were not designed to land on the Moon. Specifications for them include the following:

Materials
The fuselage was made of aluminum honeycomb sandwich panels and monolithic panels, stiffened by vertical and horizontal frames. The frames were made of metal matrix composite, which is a combination of boron fiber and aluminum tubes. The payload bay doors were made of a sandwich composite of carbon fiber–epoxy skins and a Nomex core, with carbon fiber composite stiffeners.

Windows
The windows were made of aluminum silicate glass and fused silica glass, with an internal pressure pane, an optical pane, and an external thermal pane.

Airlock
The Orbiter had an internal airlock in the mid-deck, but Discovery, Atlantis, and Endeavour had an external airlock in the payload bay [9].

Nomex was not listed in Table 6.2. It can be any of a number of organic aramid materials, all of which are highly flame-resistant. Figure 6.1 shows the generic repeating structure for Nomex. The existing process of making the material – mostly for flame-resistant clothing for first responders – is mature enough that prices for it can be found at several on-line sources.

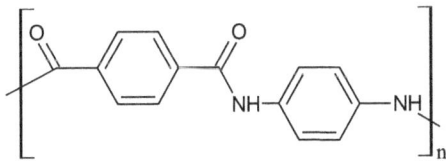

Figure 6.1: Nomex structure.

These specifications will certainly prove useful as a starting point for future craft going to the Moon and returning, and withstanding impact and landing at any landing ports. However, higher-strength alloys may be required, and improved aramid organic flame-resistant materials may also be necessary [8, 9].

References

[1] MIT Space Exploration Initiative Outreach: Selecting a Lunar Landing Site. Website. (Accessed 10 April 2025, as: https://sei-engagement.pubpub.org/pub/s2ymacig/release/8).

[2] ESA – Next generation landing technology. (Accessed 10 April 2025, as: https://www.google.com/search?q=european+space+agency%2C+Selecting+a+lunar+landing+site&sca_esv=02c44965d6d4b280&sca_upv=1&rlz=1C1GCEU_enUS819US819&ei=5UH0ZpiBM9j-ptQPoI-pgAc&ved=0ahUKEwjYh5WUyd6IAxVYv4kEHaBHCnAQ4dUDCBA&uact=5&oq=european+space+agency%2C+Selecting+a+lunar+landing+site&gs).

[3] NASA Identifies Candidate Regions for Landing Next Americans on Moon. Website. (Accessed 10 April 2025, as: https://www.nasa.gov/news-release/nasa-identifies-candidate-regions-for-landing-next-americans-on-moon/).

[4] Lunar Landing Pads KSC TOPS-89. Website. (Accessed 10 April 2025, as: https://technology.nasa.gov/patent/KSC-TOPS-89#:~:text=Lunar%20landing%20and%20launch%20pads,vehicle%20landing%20or%20take%20off).

[5] The European Space Agency, Argonaut: Europe's lunar lander programme. Website. (Accessed 10 April 2025, as: https://esa.int/Science_Exploration/Human_and_Robotic_Exploration/Exploration/Argonaut_Europe_s_lunar_lander_programme).

[6] The European Space Agency, Luna. Website. (Accessed 10 April 2025, as: https://esa.int/Science_Exploration/Human_and_Robotic_Exploration/Exploration/Luna).

[7] NSTS 1988 News Reference Manual. Website. (Accessed 10 April 2025, as: https://www.globalsecurity.org/space/library/report/1988/stsref-toc.html).

[8] Valence Surface Technologies. Website. (Accessed 10 April 2025, as: valencesurfacetech.com/the-news/aerospace-metals).

[9] NASA. Website. (Accessed 10 April 2025, as: https://ntrs.gov/api/citations/19880005632/downloads/19880005632.pdf).

Chapter 7
Mining

We mentioned in Chapter 1 that there will be several different reasons for establishing some type of permanent human presence on the Moon. A better understanding of our own planet, as well as of the Sun, is perhaps an obvious one. An understanding of how humans live and exist when not on our planet is another – which will be valuable as we attempt to reach and explore Mars with human beings, and not solely with robotic probes.

Mining materials that are valuable on Earth, and thus making some type of profit, is certainly another reason for establishing some form of long-term presence on the Moon. As an example, NASA's Lunar Reconnaissance Orbiter was launched predominantly to determine if there was usable water on the Moon's surface in areas that do not receive sunlight. It has done more than this, though, and found evidence of some precious metals in concentrated areas beneath the surface [1, 2].

7.1 What will be mined?

The Apollo missions, which landed astronauts on the Moon in the 1970s, had as a primary mission the collection of rocks from the surface and their return to Earth. This was important enough that it was one of the first activities Neil Armstrong and Buzz Aldrin undertook after disembarking from the lunar capsule. However, this collecting was intended to perform scientific studies on the materials that were gathered. In the future, we will undoubtedly need to find some economic incentive to mine materials that are on or near the lunar surface. To this end, the most recent lunar surveys include some study of the composition of the surface [3–12].

7.2 How can ilmenite be mined?

The following quote concerns the chemical composition of lunar rocks that make up the dark areas of the surface:

"Basalts are rocks solidified from molten lava. On Earth, basalts are a common type of volcanic rock and are found in places such as Hawaii. Basalts are generally dark gray in color; when one looks at the Moon in the night sky, the dark areas are basalt. The basalts found at the Apollo 11 landing site are generally similar to basalts on Earth and are composed primarily of the minerals pyroxene and plagioclase. One difference is that the Apollo 11 basalts contain much more of the element titanium than is usually found in basalts on Earth. As a result, the mineral ilmenite is abundant in Apollo 11 basalts. Another titanium-bearing mineral, armalcolite, was first discov-

https://doi.org/10.1515/9783111387307-007

ered in the Apollo 11 samples and was named for the first syllables of the last names of the three Apollo 11 astronauts. The basalts found at the Apollo 11 landing site range in age from 3.6 to 3.9 billion years and were formed from at least two chemically distinct magma sources. Prior to the lunar landings, some scientists thought that the Moon might have always been a cold, undifferentiated body. The discovery of basalt, which was once molten magma, disproved this hypothesis" [14].

The presence of significant amounts of titanium is important, since it is a valuable element on Earth. Armalcolite – named after Armstrong, Aldrin, and Collins – represents a mineral never before seen on Earth, which certainly makes one think that there may be other mineral compositions that are unique to the Moon. However, other elements may still be found in lunar basalts after more thorough study. Table 7.1 illustrates the chemical composition of the minerals just mentioned.

Table 7.1: Mineral compositions.

Name	Composition	Comments
Armalcolite	$(Mg,Fe^{2+})Ti_2O_5$	Not found on Earth; name is a combination of the astronauts' names – Armstrong, Aldrin, Collins
Basalt	Indeterminant (containing Fe, Mg, Ca, Al, K, Na)	A solid mixture
Ilmenite	$FeTiO_3$	Sometimes: $(Fe, Mg, MnTi)O_3$
Plagioclase	$(Na,Ca)(Al)(Si)O_8$	Very common in Earth's crust
Pyroxene	(Ca,Na,Fe^{2+},Mg) $(Cr,Fe^{3+},Ti)(Si,Al)_2O_6$	Common in both igneous and metamorphic rock formations

As mentioned, the lure of refining titanium from ilmenite is one that could, in theory, change the world if large quantities of either ilmenite or reduced, refined titanium can be brought back to Earth.

Titanium is an element that has proven very useful in industry as a reduced metal, with numerous niche applications. Perhaps the largest use of reduced titanium metal is the production of a limited number of submarines by the now-defunct Soviet Union. The idea of making a naval vessel from such an expensive metal became a reality because the territories of the U.S.S.R. had a significant amount of ilmenite ores, and because these submarines were supposed to be able to travel underwater more swiftly than any other submarines and dive to deeper depths.

Interestingly, titanium has also found significant and extensive use in one particular compound, titanium dioxide (TiO_2). More properly called titanium(IV) oxide, this compound has been used extensively as a white pigment, as well as an additive in paints, cosmetics, and sunscreens. Figure 7.1 shows its production from titanium compounds:

$$FeTiO_{3(impure)} + H_2SO_4 \rightarrow FeSO_4 \cdot 5H_2O_{(s)} + TiOSO_4$$

$$TiOSO_4 + H_2O \rightarrow TiO_2 \cdot xH_2O$$

$$Ti(OC_2H_5)_4 + 2H_2O \rightarrow 4H_5C_2OH + TiO_2$$

$$TiO_{2(impure)} + Cl_{2(g)}C_{(s)} \rightarrow hi\,T \rightarrow TiCl_{4(l)}$$

$$TiCl_{4(l)} + O_{2(g)} \rightarrow TiO_{2(s)}Cl_{2(g)}$$

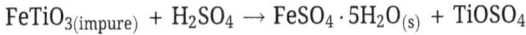

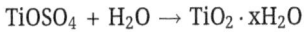

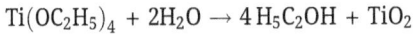

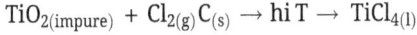

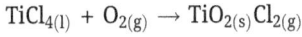

Figure 7.1: Titanium dioxide production.

Each of the reactions shown in Figure 7.1 is difficult to write as a precise, stoichiometric equation because an excess of some reagent is required. Additionally, co-reagents such as sulfuric acid, chlorine, and water are easy to find, produce, and use on Earth but will have to be imported if this reaction chemistry is to be undertaken in a lunar setting. In Chapter 3, we have already discussed the finding, purification, and use of water in any lunar habitat, and so it may be usable in this application. However, sulfuric acid and chlorine will have to be imported. Even if sulfur and sodium chloride were found in abundance on the lunar surface, the reaction chemistry to produce sulfuric acid and chlorine requires water and electricity, respectively.

Note also that these reactions do not reduce titanium to its elemental form at any point. They bring it to the useful titanium(IV) oxide. Table 7.2 lists the major uses of titanium dioxide, which is an industry that requires more than one million metric tons of the material per year. The U.S. Mineral Commodity Summaries state, "More than 95% of titanium mineral concentrates were consumed by domestic TiO_2 producers. The remainder was used in welding-rod coatings and for manufacturing carbides, chemicals, and titanium metal" [13]. This indicates that a robust market for the material exists now and probably will for the foreseeable future.

Table 7.2: Titanium dioxide usage [13].

Use	Percent of total	Comments
Paints	60	Includes paints and varnishes
Plastics	20	Additive
Paper	5	Whitening agent
Other	15	Including industrial catalysts, ceramic materials, floor coverings, inks, and roofing materials

7.3 How can helium-3 be gathered and refined?

7.3.1 Possibilities for gathering

Helium is embedded in the lunar regolith, and may be extracted by heating samples of that regolith [14–18]. This will need to be done in a heated vacuum chamber so that the released helium is not contaminated with other gases. Currently, it is believed that there are few other gases within the regolith, but further examination of the lunar crust may prove this wrong.

Helium on Earth is all helium-4, or $_2^4$He, and is found in natural gas wells. It is believed to have formed during millions of years of decay of buried matter, which gives off a certain amount of alpha particles that, over time, in turn pick up the electrons needed to create all helium-4. Figure 7.2 shows the basic reaction for this.

$$\text{alpha} + 4e^- \rightarrow {}_2^4\text{He}$$

Figure 7.2: Helium-4 production from alpha particles.

On the Moon's surface, helium is helium-3, or $_2^3$He, and is the result of helium being spalled off the Sun and carried outward by the solar wind. A small amount of this vast total from the Sun collides with, and becomes embedded on, the lunar surface.

7.3.2 Capture and return to Earth

No current system exists to transport helium-3 back to Earth from the Moon. It can be envisioned, however, that once helium is concentrated and extracted from regolith, it will be compressed into tanks, which will then be returned to Earth. The idea of compressing any gas, helium or another, is a very well-established technology and should be adaptable to a lunar operation.

Liquefying helium is a greater challenge. The gases shown in Table 7.3 are those which Apollo instrumentation has been able to detect in the regolith. It can be seen that helium has the lowest boiling point of any of these, which means it requires the most energy input to compress or liquefy.

On Earth, helium can not only be compressed in tanks, but it can, with some difficulty and care, be liquefied for either transport or storage. It is economically feasible to transport large masses of helium as a liquid, and then to distribute it to end-users in smaller amounts as tanks of compressed gas.

The cost of compression versus liquefaction will have to be analyzed when the transport of lunar helium becomes a possibility. Figure 7.3 shows the simplified schematic of a compressed gas tank, with the important features of a double wall, the traditional cylindrical shape, and an inlet/outlet valve at the top. This double wall may be necessary in a

Table 7.3: Gases found in the regolith.

Gas	Boiling point (°C)	Comments
Argon	−185.5	Deposited by solar wind; most common on this list
Carbon monoxide	−191.5	
Carbon dioxide	−78.46	
Helium	−268.9	Deposited by solar wind; lowest boiling point
Methane	−161.5	
Nitrogen	−195.8	
Oxygen	−183.0	

Figure 7.3: Schematic, compressed gas cylinder.

process that brings helium back to Earth, since an accidental puncture of a wall of the container during re-entry and landing would not mean the loss of the cargo [17–19].

It is perhaps obvious that, on Earth, these containers are cylindrical, with a valve at the top. In transport from the Moon to the Earth, any gas cylinder will have to be engineered so that there is no excessive pressure build-up at any specific portion of the tank, such as the side or the flat, bottom end. Thus, the containers may function better if they are spherical, even though this would mean a greater amount of wasted space between containers when multiple containers are shipped at the same time.

7.4 What other materials can possibly be mined?

Perhaps obviously, any metal beyond what has been mentioned here can be mined, provided sufficient quantities can be located. Lunar mining and extraction will require somewhat different techniques than those performed on Earth because of the absence of large amounts of oxygen and water, but the biggest change will be in reductive extraction and refining. On Earth, materials such as coke are available to combine with oxygen as a metal ore is reduced to the elemental metal.

Iron

The largest industrial-scale example of a reductive extraction of a metal from its ore is shown in Figure 7.4: the production of iron from iron ore in a series of reductive steps.

Using elemental hydrogen as a reducing agent

$$H_2 + 3Fe_2O_3 \rightarrow 2Fe_3O_4 + H_2O$$

$$H_2 + Fe_3O_4 \rightarrow 3FeO + H_2O$$

$$H_2 + FeO \rightarrow Fe + H_2O$$

Using carbon monoxide as a reducing agent

$$CO + 3Fe_2O_3 \rightarrow 2Fe_3O_4 + CO$$

$$CO + Fe_3O_4 \rightarrow 3FeO + CO_2$$

$$CO + FeO \rightarrow Fe + CO_2$$

Including carbon as a solid in the reaction, to produce carbon monoxide

$$CO_2 + C \rightarrow 2CO$$

Figure 7.4: Iron ore reduction to iron.

While hydrogen can be formed on the Moon through the redox of water into its elements, it is theoretically possible to use the direct reduced iron (DRI) method to produce the metal. However, even though this process operates at lower temperatures than those required for a blast furnace (the most widely used method of producing iron), it still necessitates temperatures near 1,200 °C.

Since there has been no discovery of a carbon source on the lunar surface, with the possible exception of small amounts of methane in the regolith, other possible routes for refining will most likely have to be developed. Chemical reactions will have to be developed and relied upon to introduce new reducing agents to reduce metals to their elemental form. Failing this, a significant amount of carbon will have to be imported to the Moon from the Earth. The reduction just shown, using hydrogen – the DRI method – may be a realistic option, since the production of oxygen from water, discussed in Chapter 2, will, as mentioned, produce elemental hydrogen gas as a co-product.

Rare earth elements

In the relatively recent past, NASA reports have mentioned the presence of lanthanum and neodymium in areas of the lunar regolith [20]. These elements have been detected by remote sensors from orbit, but the amounts of any particular element have not yet been estimated. The rising use of these elements in everyday life, as well as other rare earth elements, is of great interest, since such elements are finding increasing applications in modern electronics, wind turbines, electric vehicles, and smartphones.

References

[1]   NASA. Website. (Accessed 11 April 2025, as: nasa.gov/missions/radar-points-to-moon-being-more-metallic-than-researchers-thought/).

[2]   Moon richer in metals than previously thought – NASA. Mining.com. Website. (Accessed 11 April 2025, as: mining.com/moon-richer-in-metals-than-previously-thought-nasa/).

[3]   Cui, Z., et al. A sample of the Moon's far side retrieved by Change'e-6 contains 2.83-billion-year-old basalt. *Science*, 20 December 2024, 386, 6728, p.1395–1397

[4]   https://arxiv.org/ftp/arxiv/papers/2109/2109.02201.pdf . https://doi.org/10.1080/08827508.2021. 1969390 Mineral Processing and Metal Extraction on the Lunar Surface – Challenges and Opportunities

[5]   https://www.diva-portal.org/smash/get/diva2:1664757/FULLTEXT01.pdf A concept design to enable lunar mining

[6]   https://www.sciencedirect.com/science/article/pii/S0094576522005021 Acta Astronautica, Defining the notion of mining, extraction and collection: A step toward a sustainable use of lunar resources

[7]   https://www.science.org/content/article/moon-s-scientifically-important-sites-could-be-lost-forever-mining-rush Defining the notion of mining, extraction and collection: A step toward a sustainable use of lunar resources

[8]   *SPACE: SCIENCE & TECHNOLOGY*, 1 Jun 2023, Vol 3, Article ID: 0037, DOI: 10.34133/space.00373. https://spj.science.org/doi/10.34133/space.0037. Overview of the Lunar In Situ Resource Utilization Techniques for Future Lunar Missions

[9]   Understanding the Lunar Surface and Space-Moon Interactions, *Reviews in Mineralogy & Geochemistry* Vol. 60, pp. 83–219, 2006, https://www.higp.hawaii.edu/~gjtaylor/GG-673/Moon/New%20Views/Chapter_2_Lucey%20et%20al%202006%20Understanding%20the%20Lunar%20Surface%20and%20Space%20Moon%20Interactions.pdf.

[10]   The Lunar Gold Rush: How Moon Mining Could Work, https://www.jpl.nasa.gov/infographics/the-lunar-gold-rush-how-moon-mining-could-work

[11]   NASA sees moon lunar trial mining in the next decade, https://www.reuters.com/science/nasa-sees-moon-lunar-mining-trial-within-next-decade-2023-06-28/

[12]   LPI Resources, Apollo 11 Lunar Samples. Website. (Accessed 25 September 2024, as: https://www.lpi.usra.edu/lunar/missions/apollo/apollo_11/samples/).

[13]   USGS Mineral Commodity Summaries 2023, downloadable.

[14]   Feasibility of lunar Helium-3 mining, https://ui.adsabs.harvard.edu/abs/2014cosp. . .40E1515K/abstract

[15]   European Space Agency, Helium-3 Mining on the Lunar Surface. Website. (Accessed 25 September 2024, as: https://www.esa.int/Enabling_Support/Preparing_for_the_Future/Space_for_Earth/Energy/Helium-3_mining_on_the_lunar_surface).

[16] Harnessing Power from the Moon. Website. (Accessed 25 September 2024, as: https://www.nasa.gov/directorates/stmd/space-tech-research-grants/harnessing-power-from-the-moon/).

[17] National Academy Press. The Helium Supply Chain. Website. (Accessed 25 September 2024, as: https://nap.nationalacademies.org/read/12844/chapter/5).

[18] Chart Industries. Liquid Helium Containers. Website. (Accessed 25 September 2024, as: chrome-extension://efaidnbmnnnibpcajpcglclefindmkaj/https://files.chartindustries.com/10576741_Ultra_Helium_Dewars_Product_Manual_Rev_A_ws.pdf).

[19] Lunar Surface Science Workshop 18 (2022). Understanding the economic worth of precious lunar metals. John C. Johnson, Peter A. Johnson, Austin A. Mardon. Website. (Accessed 25 September 2024, as: chrome-extension://efaidnbmnnnibpcajpcglclefindmkaj/https://www.hou.usra.edu/meetings/lunarsurface18/pdf/6001.pdf).

[20] NASA: Extralunar Materials in Lunar Regolith. Website. (Accessed 11 April 2025, as: https://solarsystem.nasa.gov/studies/56/extralunar-materials-in-lunar-regolith).

Chapter 8
Vehicles

Moving about on the lunar surface can be as straightforward as walking in suits, as was done during all the Apollo missions that placed one or more individuals on the Moon. A space suit may not seem like a vehicle in terms with which we are familiar, yet such a suit is an essential means of moving about when some larger vehicle is not present or is not considered necessary, since the Moon has no atmosphere. Each suit is equipped with the necessary systems to allow a person to breathe. The Apollo suits contained 7–8 h of air, with the difference being related to how hard the astronaut was working.

The technology certainly exists to construct suits that can allow a person to walk on the Moon, and what can be produced now is longer-lasting than what was used in the 1960s and 1970s. However, it should be remembered that the lunar suits worn by Apollo mission astronauts functioned well for the extended time during which the astronauts were out of their capsule and on the lunar surface – short periods within a span of a few days. Figure 8.1 shows a simple schematic of the layers successfully used in a lunar suit.

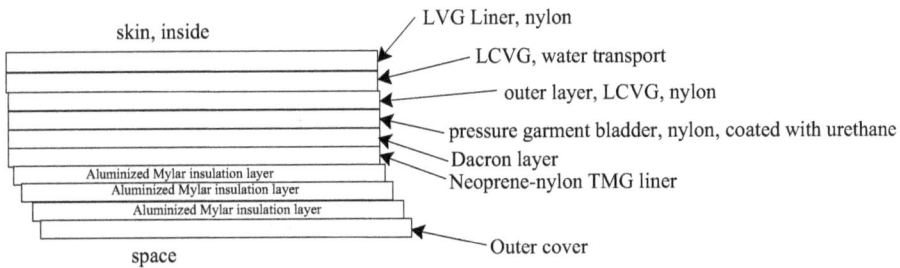

Figure 8.1: Cross-section of a lunar suit.
Nylon
Polyurethane
Dacron – polyester
Neoprene – polychloroprene, a synthetic rubber developed in 1930, DuPont
LCVG – liquid cooling and ventilation garment, usually made of polyvinyl chloride (PVC)

The most modern space suits – known as Extravehicular Mobility Units (EMU) – are still designed to function for approximately seven hours. The suits carry both oxygen and a carbon dioxide removal system in what is called a portable life support system. Oxygen is provided via two tanks worn on the astronaut's back. However, it is the effort expended by the astronaut that tends to limit the length of use of the EMU (in short, the work can be hard and physically challenging). The suits also carry water for the user to drink, as seven hours of labor without any rehydration can be dangerous

https://doi.org/10.1515/9783111387307-008

[1–4]. Additionally, the suits weigh 113 kg (which is not overly high for lunar gravity), including the life support backpack [1]. Suits being designed by the European Space Agency (ESA) generally have considerable similarities [5–6].

The removal of carbon dioxide that the astronaut has breathed into the EMU is a chemical reaction that turns the gas into a solid utilizing lithium hydroxide. Figure 8.2 shows the basic chemistry. This is essentially the same reaction that is used as a back-up in submarines to remove excess CO_2 from the boat (the primary system in a submarine utilizes monoethanolamine). As long as there is moisture in the air, the lithium hydroxide will react with the CO_2 since it is highly deliquescent. Water is provided in the suit in what is essentially a plastic bag located in the helmet of the suit.

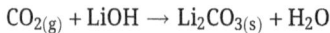

$$CO_{2(g)} + LiOH \rightarrow Li_2CO_{3(s)} + H_2O$$

Figure 8.2: Carbon dioxide scrubbing from an EMU.

When ore deposits of significant value are located by any resident of a lunar colony, it may prove far more economically feasible to employ some form of wheeled vehicle or light rail to move products from the mine to a landing port, to transport material and equipment to the mine site from a port, and to move personnel in both directions. Thus, an examination of previous lunar rovers, as well as existing light rail systems on Earth, is warranted.

8.1 What wheeled or tracked vehicles will be used?

The only vehicles to have explored any area of the Moon are the lunar rovers from the early 1970s and the Apollo missions. These functioned using a series of silver-oxide batteries and had a range of several kilometers. This distance was not tested, though, because the astronauts never went farther than half the distance that their suits could travel, which was significantly less than the range of the rovers. Thus, if there was any problem with the rover, they would still be able to return to the module so they could safely return to the Apollo capsule still in orbit. Figure 8.3 shows one photo of a lunar rover [6].

It is logical to assume that in any future lunar habitat, the designs of the lunar rovers may, in some part, be "dusted off" and reused. Before constructing more than the minimal number of new rovers, however, it might be wise to visit those which have been on the Moon since the 1970s and examine them to determine what sort of degradation has occurred in the decades during which they have remained on the Moon. For example, have the tires, which were made specifically for these rovers, undergone degradation after decades of exposure on the lunar surface? These were specially designed by Goodyear using a titanium tread over a zinc-coated wire base.

Figure 8.3: Apollo program lunar rover.
[Photo courtesy of NASA]

Have the exposed metal parts suffered any degradation over time? Have the silver oxide batteries degraded enough that they cannot be recharged?

The reaction chemistry for a silver-oxide battery is shown in Figure 8.4.

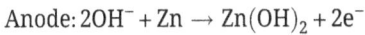

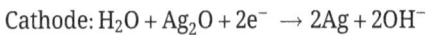

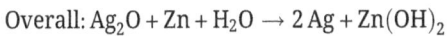

Anode: $2OH^- + Zn \rightarrow Zn(OH)_2 + 2e^-$

Cathode: $H_2O + Ag_2O + 2e^- \rightarrow 2Ag + 2OH^-$

Overall: $Ag_2O + Zn + H_2O \rightarrow 2\,Ag + Zn(OH)_2$

Figure 8.4: Silver-oxide battery couple.

It is noteworthy that water is part of the charge and discharge reaction for a silver-oxide battery. While batteries are in sealed packets or containers, certainly for the lunar rovers, there is no data that exists concerning how water would react, or evaporate, when in such containers in an extreme environment for nearly fifty years.

Perhaps the most recent vehicles that have demonstrated the ability to endure harsh environments beyond Earth are the twin robots Opportunity and Spirit, on Mars. These two were designed to function for a period of months, yet Spirit was operational from 2004 to 2010, while Opportunity functioned from 2004 to 2018. Both had wheels that were treaded heavily enough to be very close to tank treads, as opposed to traditional wheels. Both utilized lithium-ion batteries, which were charged by solar cells. Figure 8.5 shows the lithium-ion battery couple.

It should be noted that the loss of lithium – or rather, its movement between graphite and a cobalt species – is not stoichiometric in the initial charging sequence, nor is it stoichiometric in the subsequent discharge. Thus, the letters "x" and "y" are used to indicate incomplete movement.

Initial charge: $C_6 + LiCoO_2 \rightarrow Li_{1-x}CoO_2 + Li_xC$

Discharge: $Li_{1-x}CoO_2 + Li_xC_6 \rightarrow Li_{1-x+y}CoO_2 + Li_{x-y}C_6$

Figure 8.5: Lithium-ion battery charging and discharging.

But the greatest advantage of lithium-ion batteries on Mars rovers will be the same as that for any lunar vehicles: they are lightweight – as is evident in Figure 8.6. Once again, the transport of all items to a lunar habitat and colony will depend on the weight lifted to the Moon.

The solar panels used on Opportunity and Spirit are triple-junction solar panels made from gallium arsenide. Since this is an established type of solar panel, it can be assumed that the same might be useful on any vehicles to be used in a future lunar habitat and vehicle.

Figure 8.6: Mars rover, showing solar panels.
[Photo courtesy of NASA]

The current plans for vehicles to be used in any future presence on the Moon have been dubbed lunar terrain vehicles, or LTVs, by NASA. These appear to be something of a hybrid between the older vehicles used in the Apollo missions and those used on Mars [6–10].

NASA describes the LTV as follows:

"Human mobility on the lunar surface is crucial for enhancing scientific discovery on each mission and preparing for planetary mobility on Mars. Instead of owning the vehicle, NASA plans to contract it as a service from industry. The LTV will be the ultimate lunar surface terrain vehicle with advanced power management, autonomous driving, state-of-the-art communications and navigation systems, along with other extreme environment technologies that will provide the ability to collect and conduct science while keeping astronauts and the vehicle safe and ready for its next mission" [6].

It can be implied from this broad description that tires will again be of great importance and may once more use a metal composite for maximum traction and durability. Additionally, the statement suggests that the LTV will be versatile enough to function either robotically or with a crew. The mention of "state-of-the-art communications" implies that an onboard computer will be present, which will most likely need to be shielded from direct solar radiation.

The ESA is also developing rovers for use on the Moon, including what has been titled the Argonaut Lunar Lander, the Lunar View, the Zebro rover (a six-legged rover), and the European Moon Rover Systems. The Argonaut should be able to carry up to 1,500 kg of payload and should be able to function with a crew or autonomously [11–13].

8.2 How will light rail be utilized?

On Earth, modern light rail tends to be a mode of transportation that is reserved for cities or areas in the immediate vicinity of cities. Curiously, for some time at the end of the nineteenth and the beginning of the twentieth centuries, the light rail line established in Namibia – what was then German Southwest Africa – was the longest narrow-gauge rail in the world and a means of transporting workers and diamonds from the fields inland to the seacoast. Thus, light rail has a significant history, even if it is not particularly well remembered.

The concept of any type of rail for use in a lunar colony will probably only be implemented after significant exploration has been conducted and some extractable mineral has been located in large quantities. It is simply far less expensive to utilize some kind of wheeled or tracked vehicle (basically a truck or backhoe) than it is to install rails and import rail cars of any sort. Still, the following considerations can be made:

For any lunar colony, since the gravity is 1/6th that of Earth, it is quite probable that light rail could actually haul heavy materials, such as ore. The other factor that makes light rail an attractive option on the lunar surface is the cost of bringing any materials, be they tracks or rail cars, to the Moon. Rail should cost more at the outset but may cost significantly less with the passage of time [14–15].

A scenario for lunar light rail might include:
1. Rails made of aluminum.
2. Electric train motor.
3. Cars made solely as flatbed cars, with tie-downs or inflatable polymer bubbles atop the flatbed if needed.
4. Solar-powered electric charging stations along the rail routes.

An explanation of each, and of the materials chemistry involved in each of them, includes:
1. By making the rails of aluminum or magnesium-aluminum alloy, they will have as light a mass as possible. These will have to be refined and formed on Earth and transported to any lunar base. We currently have no indication of bauxite deposits in any area of the lunar surface from which aluminum could be refined.
2. Electric motors will be essential to any light rail that functions on the lunar surface, for several reasons:
    a. Electric motors can be recharged via energy harvested from solar power.
    b. Electric motors do not require an external, liquid fuel.
    c. Electric motors do not necessarily require air to function – although some use it as a coolant.
3. Flatbed cars are the simplest form of rail car. This is important since they will need to be imported from Earth. Additionally, a flat surface – the flatbed – can be modified to carry ores, equipment, or people, much like different units are currently attached to semi-trailer trucks. The idea of an inflatable habitat that can be attached to a flatbed car is much the same as a living module that could be used in any stationary lunar construction – somewhat like the British Antarctic habitat mentioned in Chapter 5. The plastic "skin" of any such inflatable habitat must be as robust as any other area in which people will reside, even if only for a short period.

Clearly, the idea of some rail line on the lunar surface remains an idea for a more distant future, after some human presence on the Moon has become an established fact. It can be argued, though, that it remains worthy of discussion at this point, both in terms of traditional rail and in terms of the more cutting-edge mag-lev rail design.

8.3 Will mag-lev trains be a viable option?

One of the holy grails of rail transport on Earth is that of a magnetic levitation train – often called a maglev – that can travel great distances while expending almost no energy. Such conservation of energy exists because sets of magnets keep the train and track apart, thus keeping the resistance to movement at zero (all other trains have some level of resistance between the wheels and the tracks). This has been achieved on Earth in only a few projects that have reached commercial application, although extensive and significant research efforts continue. Companies heavily invested in maglev trains include those listed in Table 8.1. Additionally, there are now organizations that advocate for mag-lev trains as a possibility within the greater sphere of high-speed rail, often to link cities that have large swaths of rural areas between them [16–20].

Table 8.1: Maglev train companies [21–28].

Company name	Location	Website
Alstom	France	https://www.alstom.com
American Maglev (AMT)	Florida, United States	http://american-maglev.com
CRRC Qingdao Sifang Corp. Ltd.	China	https://www.crrcgc.cc/sfgfen
SCMAGLEV, Central Japan Railway Company	Japan	https://scmaglev.jr-central-global.com
Hyundai Rotem Company	South Korea	https://www.hyundaimotorgroup.com
Mitsubishi Heavy Industries, Ltd.		https://www.mihi.com
Shanghai Maglev Transportation Development Co., Ltd.	China	http://www.smtdc.com/en
SwissRapide AG	Switzerland	https://www.swissrapide.com

Once again, a perhaps obvious advantage to the idea of maglev trains on the Moon is the much lower gravity – roughly 1/6th that of Earth. In theory, it should be easier to keep any train moving since there is much less attractive force between any train car and the track. Figure 8.7 shows the basic schematic of a mag-lev train and the separation of the train and the track.

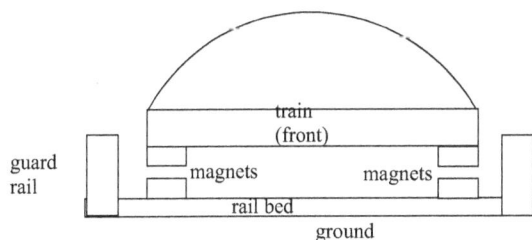

Figure 8.7: Mag-lev train schematic.

Finally, the idea of some kind of lunar rail is an enticing one, but it will also be the most expensive of any surface transport system that can be initiated and constructed.

What will be simpler, even if it sounds almost childish, is to have a designated area on which wheeled or tracked vehicles can travel. This would be ecologically advantageous, since running wheeled vehicles over the same areas repeatedly will minimize introduced human interaction and any possible damage to the surface.

Importantly, precedent for the controlled movement of large vehicles over valuable terrain already exists. Large tractors operated by farmers in the United States currently run on designated paths in large fields, so that only a minimal amount of arable land is overrun by such vehicles, and a maximal amount of land is farmed. Such tractors have been guided by satellites, ensuring that their paths minimize any trampling of growing crops. It is conceivable that such a system could be put in place for any lunar colony as well.

References

[1]   NASA Extravehicular Mobility Units (EMU). Website. (Accessed 12 April 2025, as: https://nasa.gov/suits-and-rovers).

[2]   Extravehicular Mobility Unit. Website. (Accessed 12 April 2025, as: https://nasa.gov/wp-content/uploads/2015/07/2019_07_ea_emu.pdf).

[3]   The Space Shuttle Extravehicular Mobility Unit. Website. (Accessed 12 April 2025, as: https://nasa.gov/wp-conntent/uploads/2009/07/188963main_extravehicular_mobility_unit.pdf?emrc=2aa548).

[4]   Spacesuits. Website. (Accessed 12 April 2025, as: https://nasa.gov/humans-in-space/astronauts/spacesuits/).

[5]   Winning spacesuit designs. Website. (Accessed 12 April 2025, as: https://esa.int/Science_Exploration/Human_and_Robotic_Exploration/Winning_spacesuit_designs).

[6]   ESA, Human and Robotic Exploration. Website. (Accessed 12 April 2025, as: https://esa.int/Science_Exploration/Human_and_Robotic_Exploration).

[7]   The Apollo Lunar Roving Vehicle, downloadable at: nssdc.gsfc.gov/planetary/lunar/Apollo_lrv.html

[8]   NASA, The Development of Wheels for the Lunar Roving Vehicle. Website. (Accessed 12 April 2025, as: https://ntrs.nasa.gov/api/citations/20100000019/downloads/20100000019.pdf).

[9]   NASA Selects Companies to Advance Moon Mobility for Artemis Missions. Website. (Accessed 12 April 2025, as: https://nasa.gov/news-release/nasa-selectscompanies-to-advance-moon-mobility-for-artemis-missions/).

[10] NASA. Website. (Accessed 12 April 2025, as: nasa.gov/suits-and-rovers/lunar-terrain-vehicle).

[11] ESA. Argonaut: Europe's lunar lander programme. Website. (Accessed 11 April 2025, as: https://esa.int/Science_Exploration/Human_and_robotic_Exploration/Exploration/Argonaut_Europe_s_lunar_lander_programme).

[12] Universe Today. ESA is Testing a Modular Multipurpose Rover that Could Be a Science Lab or a Tiny Bulldozer. Website. (Accessed 12 April 2025, as: universetoday.com/articles/esa-is-testing-a-modular-multipurpose-rover-that-could-be-a-science-lab-or-a-tiny-bulldozer).

[13] ESA-Moon rover. Website. (Accessed 12 April 2025, as: https://esa.int/ESA_Multimedia/Keywords/Description/Moon_rover/(result_type)/videos).

[14] LRTA. Light Rail Transit Association. Website. (Accessed 12 April 2025, as: https://www.lrta.info).

[15] European Light Rail Conference. Website. (Accessed 12 April 2025, as: https://rail-research.europa.eu/calendar/eu-light-rail-conference).

[16] The International Maglev Board. Website. (Accessed 12 April 2025, as: https://maglevboard.net/en).

[17] Europe's Rail. Europe's Rail Joint Undertaking. Website. (Accessed 12 April 2025, as: https://euro pean-union.europa.eu/institutions-law-budget/institutions-and-bodies/search-all-eu-institutions-and -bodies/europes-rail-joint-undertaking_en).

[18] MaDe4Rail – Exploring non-traditional and emerging maglev-derived systems. Website. (Accessed 12 April 2025, as: https://rail-research.europa.eu/latest-news/made4rail-exploring-non-traditional-and-emerging-maglev-derived-systems).

[19] IHRA. International High-Speed Rail Association. Website. (Accessed 12 April 2025, as: https://ihra-hsr.org/en/organization/).

[20] Australian High Speed Rail Association. Website. (Accessed 12 April 2025, as: https://www. auhsr.org).

[21] Alstom. Website. (Accessed 12 April 2025, as: https://www.alstom.com).

[22] American Maglev (AMT). Website. (Accessed 12 April 2025, as: https://american-maglev.com).

[23] CRRC Qingdao Sifang Corp Ltd. Website. (Accessed 12 April 2025, as: https://www.crrcg.cc/sfgfen).

[24] SCMAGLEV, Central Japan Railway Company. Website. (Accessed 12 April 2025, as: https://scmaglev. jr-central-globl.com).

[25] Hyundai Rotem Company. Website. (Accessed 12 April 2025, as: https://www.hyundaimo torgroup.com).

[26] Mitsubishi Heavy Industries, Ltd. Website. (Accessed 12 April 2025, as: https://mihi.com).

[27] Shanghai Maglev Transportation Development Co., Ltd. Website. (Accessed 12 April 2025, as: https://smtdc.com/en).

[28] SwissRapide AG Website. (Accessed 12 April 2025, as: https://www.swissrapidecom).

Chapter 9
Power

It is perhaps obvious that in any lunar colony or habitat, power will be critically important. On Earth, roughly three-quarters of all power is generated by the combustion of
coal or oil, options simply not available on the Moon. Likewise, we know of no uranium
or thorium that can be refined on or near the lunar surface, so the option of utilizing
nuclear power in any way that we would use it on Earth does not exist – although the
possibility of importing nuclear material to create and build a nuclear power plant is
not theoretically impossible. On Earth, power is also generated by hydroelectric dams –
another impossibility on the Moon. Likewise, an increasing amount of power is generated by wind – a further impossibility for any lunar base or colony.

Having just removed several known, conventional sources of power as possibilities in any lunar habitat, we arrive at sources of renewable energy that should be feasible. As when we previously examined non-renewable power sources, we need to list
and evaluate renewable sources as they exist, and determine their feasibility on the
Moon. Several potential sources of renewable, clean energy, and thus of power, do
exist. These include solar power, either from photovoltaics or concentrated solar, hydrogen power, pumped water within constructed spaces, and energy harvesting. We
will examine these in this chapter.

9.1 What is the best solar power scenario?

The sun shines on some areas of the Moon at all times, making the possibility of solar
power for any lunar habitat and operations related to such a habitat a viable option.
A large number of organizations exist to advocate for solar power [1–10], and several
companies are in the business of selling electricity generated through solar power
equipment of one design or another [11–19].

The general public has become accustomed to solar panels on buildings or road-
sides, or perhaps in smaller applications, such as on individual homes. Current solar
power panels on Earth work using photovoltaic generation of electricity, by turning the
sun's rays into electricity. Figure 9.1 illustrates the basic design of a photovoltaic cell.

Existing solar cells are produced through a series of chemical reactions, usually
redox reactions, since the materials are all mined, purified, and brought together in
layers. For example, gallium and arsenic, used as gallium arsenide in solar cells, are
both refined as secondary products from other mining operations. Additionally, dopants like zinc or aluminum are also required to enhance performance and must be
mined and refined.

There are no gallium mines, for instance, but gallium is extracted as a by-product
of bauxite refining to aluminum. Figure 9.2 illustrates the basic chemical steps of this:

https://doi.org/10.1515/9783111387307-009

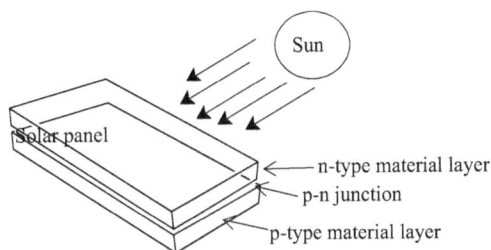

Figure 9.1: Photovoltaic electricity generation.

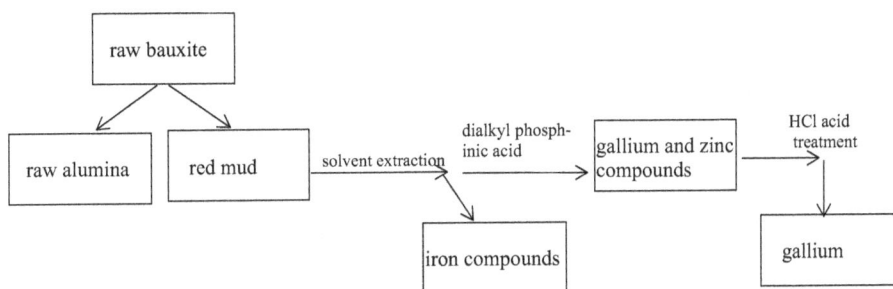

Figure 9.2: Gallium refining steps.

It is difficult to write stoichiometric reactions for the refining of metals like gallium since different ore batches will always contain varying amounts of the various metals. Of note, we can see that:

1. A bauxite by-product is "red mud."
2. Solvent extraction, which is used to remove metals such as iron, essentially rids the material of its red color.
3. Solvents such as the just-mentioned dialkyl phosphinic acid extract gallium and zinc (in the past, this has been marketed as Cyanex 272).
4. Various acid treatments, using HCl, H_2SO_4, and possibly $H_2C_2O_4$ are used for further metal separation.

The reduction of arsenic from ores undergoes a similar set of treatments. The following list shows the basic, broad steps by which arsenic is refined from arsenic trioxide. As with gallium, it is difficult to write stoichiometric chemical reactions for these processes because of differing amounts of each element in the starting material:

1. Crude As_2O_3 is gathered as a by-product of metal smelting, especially from copper, lead, and cobalt ores. Because of the toxicity of arsenic and its compounds, strict safety precautions, including the use of full-body PPE, are required for this and all steps in arsenic purification.
2. The material is physically reduced to a powder.

3. Roasting – this heating process drives off low-boiling materials and concentrates As_2O_3.
4. Repeated sublimation – the concentrated material is heated to vapor, condensed back into solid form, and the process is repeated as needed.

The production of solar panels is an exacting task on Earth, involving precise additions of gallium and arsenic, plus necessary dopants. Additionally, quality control and assurance of the synthesis occur post-production of the GaAs wafers.

It appears obvious that for any nascent lunar station or colony, panels will have to be constructed on Earth, made in small enough pieces so that they can be part of any payload to be lifted to the Moon, then assembled at the lunar site of choice – even if the raw materials are found somewhere on the Moon's surface. Existing solar cells within solar panels generally function as shown in Figure 9.3, using different layers of semi-conducting elements.

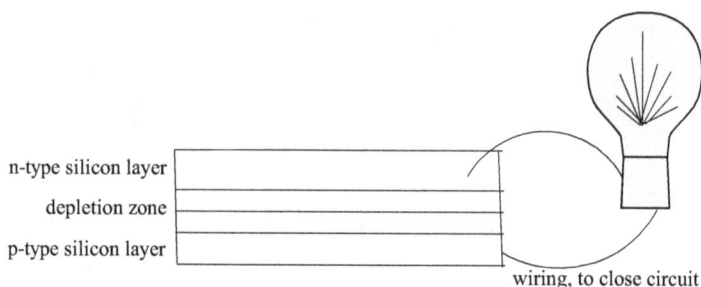

n-type silicon layer
depletion zone
p-type silicon layer
wiring, to close circuit

Figure 9.3: Basic schematic of a solar cell.

Numerous solar cells, like those in Figure 9.3, are connected to create various forms of solar panels and the very large arrays that can be seen in sunny areas on Earth. The general public is often familiar with such arrays, either seen spread across fields or sometimes on building rooftops. In the United States, these have become quite common in western states where deserts are the major land feature, but they are now being seen in non-desert areas as well. Figure 9.4 shows an array of photovoltaic solar cells.

It is sometimes easiest to understand how solar cells work by describing their function in steps, as follows:

1. Cells are always composed of two semi-conductor layers. Very often, these are silicon-based and are called an n-type and a p-type layer.
2. When silicon is used, phosphorus or arsenic in small amounts is used to dope the silicon in the n-type layer. Each of these has an extra electron compared to silicon.
3. In a similar manner, the p-type layer can be silicon doped with a precise, small amount of an element that is electron-deficient in comparison, such as gallium, indium, or boron. Such elements are said to provide what are called positively charged "holes."

4. Electrons from the n-layer can routinely move into the holes in the p-layer when they are placed adjacent to each other. This type of electron movement from the n to the p-layer results in what is called a "depletion zone." Electrons fill holes in this zone.
5. Upon the filling of the depletion zone, the n-layer will now possess cations (positive charges), while at the same time, the p-layer possesses anions (negative charges) – which creates an electric field.
6. When the PV material is irradiated with sunshine, it ejects electrons from the silicon, creating holes – which subsequently shift the electrons from the p-layer to the n-layer, thus re-creating holes in the p-type layer.
7. The photoelectric effect is essentially the release of electrons due to the sun shining on this material.
8. When the p-layer and the n-layer are linked with a conductor, often a metal wire, electrons move from the n-layer to the p-layer and subsequently return to the n-layer via the wire. This is electrical flow, or current.

Table 9.1 shows that elements such as gallium and arsenic, which are required for both the p-doped and n-doped layers, are elements with sources and locations that are well-known and established on Earth, but are, as of yet, not known in any area of the Moon. It is thus obvious that it will take future discoveries for us to examine in detail the possibility of refining and constructing such solar panels at any lunar site, assuming the raw materials can be found.

Table 9.1: Arsenic and gallium locations [20].

Element	Location	Comment
Arsenic	China	24 K metric tons (est)
	Morocco	
	Peru	27 K metric tons (est)
	Philippine	
	USA	Only at a copper smelter in Washington State
Gallium	Canada	
	China	Leading producer, 423 metric tons
	Japan	
	Russia	
	USA	

Figure 9.4: Array of photovoltaic solar cells.
[Photo courtesy of National Renewable Energy Laboratory]

9.1.1 Concentrated solar power (CSP) possibilities

The other broad and established means by which solar power is harvested on Earth is CSP.

It may seem that the idea of a power generation method requiring the flow of a liquid such as water is ridiculous for the Moon. However, we need to take into consideration that water has been found near the lunar surface, although it is not yet known how large the deposits are. Of note, CSP on Earth is often located in desert areas, where water must be conserved, kept in strictly closed loops, and brought in from significant distances by some piping system. The same engineering can be utilized in any lunar CSP arrays, retaining 100% of the water incorporated in any loop, at least in theory.

Additionally, water in any lunar colony will most likely have to be heated for a variety of reasons, such as cooking, bathing, or some industrial processes. Depending on the design, CSP heats the liquid that flows through it, making it useful both as steam to turn a turbine and as a hot liquid for other uses.

Existing CSP designs span a wide variety [11–19]. Figure 9.5 shows mirrored arrays surrounding a central point, which is the fluid-containing target for solar radiation. Figure 9.6 illustrates what is called a linear array of CSP, with mirrors surrounding a central pipeline. Perhaps obviously, in both arrangements, water must be in a liquid state and is either heated while still remaining a liquid or heated to the point of boiling. It is difficult to predict which technique will be easier to use in a lunar environment.

Figure 9.5: CSP around a central point.
[Photo courtesy of NREL]

Figure 9.6: CSP in a linear array.
[Photo courtesy of NREL]

9.1.2 Use of water for CSP, then a secondary use

When water is mined on the Moon and either stored as ice or melted and transported via tubes, it can be envisioned that the liquid can be run through tubing as warm water, and this can be used as a means of heat transfer. The purification of such water is the same as that described in Chapter 3, but its use as a CSP fluid can become the first in what will be at least a two-tiered set of uses. For example:
1. First use: generation of electricity via boiling water turning turbines.
2. Second use:

    a.   Hot water for cooking.

    b.   Heated water for bathing.

    c.   Water for dissolving some powdered mineral.

3.   Recycling water that was not used for further re-heating.

9.2 How can hydrogen power be utilized?

We have already looked at the chemistry of splitting water into elemental oxygen and hydrogen in Chapter 3. The basic chemistry is shown again in Figure 9.7.

$$2H_2O_{(l)} \rightarrow 2H_{2(g)} + O_{2(g)}$$

Figure 9.7: Separation of water into its elements.

Hydrogen has become a potential secondary source of automotive fuel on Earth, but it seems more likely that hydrogen generated on the Moon will be used as part of some rocket fuel mix. In an interesting twist, this requires combining with some element – on Earth, this is oxygen.

    This circular use becomes a challenge in terms of how useful a fuel hydrogen can be, since it is split from water to begin with. Simply recombining it with oxygen to, in some way, generate power is a closed loop and cannot generate more power than it uses. Figure 9.8 shows the problem.

$$2H_2O_{(l)} \rightarrow 2H_{2(g)} + O_{2(g)} \rightarrow \text{followed by} \rightarrow 2H_{2(g)} + O_{2(g)} \rightarrow 2H_2O_{(g) \text{ or } (l)}$$

Figure 9.8: Hydrogen generation, followed by re-use.

Some further step or source of energy must be inserted into this system. In any lunar operation, it is conceivable that the energy generated from either PV or CSP solar power will be sufficient to make this loop a workable one, as opposed to one that loses energy at some step.

9.3 What role will batteries play?

Initially, all batteries will have to be imported from Earth, simply because the location, mining, and refining of materials will be difficult. Battery manufacturing is already a highly established industry, with numerous companies producing different types of batteries, and national and international organizations monitoring and advocating for their use [21–36].

Batteries can be connected to any source of electricity, ultimately for the storage of energy for later use. Lead-acid batteries are firmly established and, ironically, the least expensive, but they are heavy. As we have mentioned before, the weight of any object brought to the Moon, including batteries, will be monitored, with the aim being the reduction of weight wherever possible.

In the past few decades, a great deal of research and development has been aimed at the production of lithium-based batteries, including lithium-air and lithium-polymer batteries, simply because they are the lightest.

It is far easier to envision batteries that have been brought to the Moon via rocket than it is to imagine manufacturing them on the Moon, in any sort of lunar workshop. The existing example is the lunar rovers that were brought to the Moon in the 1970s, and that are still on the surface. These used silver-oxide batteries. The reaction chemistry for this electrochemical couple is shown in Figure 9.9.

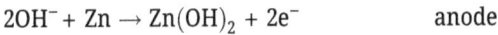

$$Ag_2O + H_2O + 2e^- \rightarrow 2\,Ag + 2OH^- \qquad \text{cathode}$$

$$2OH^- + Zn \rightarrow Zn(OH)_2 + 2e^- \qquad \text{anode}$$

Figure 9.9: Silver-oxide battery couples.

Like most electrochemical couples, a small amount of water is required for electron transfer to occur. This, in turn, means that a future requirement will be that all such batteries are encased with no leakage. This is common in existing batteries, but the casings may need to be made of some enhanced, reinforced material in a lunar environment, where even a small leakage could be catastrophic.

There is certainly a case to be made for utilizing some type of lithium battery in as many lunar habitat situations as possible, simply because lithium is the least dense of the metals. What is called a lithium metal battery utilizes the reaction chemistry shown in Figure 9.10.

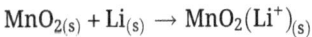

$$MnO_{2(s)} + Li_{(s)} \rightarrow MnO_2(Li^+)_{(s)}$$

Written as half cells:

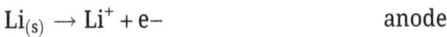

$$Li^+ + MnO_{2(s)} + e- \rightarrow MnO_2(Li^+) \qquad \text{cathode}$$

$$Li_{(s)} \rightarrow Li^+ + e- \qquad \text{anode}$$

Figure 9.10: Lithium-metal battery.

While this cell is straightforward, it is a primary one, meaning a cell that simply discharges. Batteries that can be recharged repeatedly are referred to as secondary cells. Also, lithium-based cells require dimethyoxyethane or propylene carbonate as an electrolyte. Using them in some lunar colony will require a means by which the material can be either recharged or reused – in case some of the solvent leaks from the cell. Aside from their lighter weight, of interest is most likely the high voltage that can be generated: 3.4 V.

Figure 9.11 illustrates the reaction chemistry involved in what is termed the lithium-ion battery couple.

$$C_6 + LiCoO_2 \rightarrow Li_{1-x}CoO_2 + Li_xC_6 \qquad \text{initial charging}$$

$$Li_xC_6 + Li_{1-x}CoO_2 \rightarrow Li_{1-x+y}CoO_2 + Li_{x-y}C_6 \qquad \text{discharge}$$

Figure 9.11: Lithium-ion battery reactions.

Note that in Figure 9.11, the initial charge intercalates some lithium into the graphite that is present. The discharge illustrates that some, but not all, of the intercalated lithium returns to the lithium-cobalt-oxide, which is the initial source. This is why a solvent must be part of the cell system.

9.4 Can pumped water be used to generate power?

The term "water power" has such deep, easy-to-imagine imagery on Earth that using it in relation to any lunar habitat might seem absolutely laughable. However, a future possibility exists that has never needed to be explored on Earth. Water that is pumped or gravity-fed from one place to another can have small turbines placed strategically in any pipe along a pathway, which could, in theory, be used to generate electricity.

A schematic of a hypothetical water flow and power harvesting from that flow is shown in Figure 9.12.

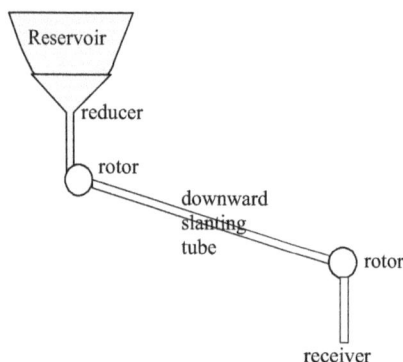

Figure 9.12: Power generation from water flow.

Water can still be fed by gravity from a source – a holding tank or reservoir – to further points, be they end points where someone will use them directly, or some holding tank farther down the line. Between the start point and any down-the-line point, small turbines can be installed at the edges of the pipes, as shown, and they turn whenever water flows. Figure 9.12 shows two, but in theory, there can be considerably more, especially if the downward slope is significant. The amount of energy produced by a single turbine may be small, but if enough are installed at multiple points in any water line, the cumulative amount of electricity generated has the potential to become much larger.

In a similar manner, it is conceivable that ice can be mined from some spot, then transported and stored in an elevated position. This is much like how municipal water towers store water today. However, in a lunar scenario, the ice can be melted through the positioning of the tower in relation to areas that receive constant sunlight, allowing the ice to melt. The resultant water flows downward, and, once again, small turbines can be placed in any water-carrying pipes, resulting in some energy generation via the flowing water. Figure 9.12 essentially shows the basic schematic of this, minus only the mirror or mirrors that would melt any ice supply.

9.5 Can energy harvesting be utilized?

The above scenario of extracting energy from relatively small amounts of flowing water is a step in the direction of what is generally called energy harvesting, also termed energy scavenging or power harvesting. The idea is that, through human movement or the repeated movement of some device, energy which is normally lost can, in some way, be gathered and often stored in devices such as batteries. In any lunar habitat, the gathering and use of even small amounts of energy have the potential to be important and useful.

A significant amount has been written about energy harvesting, once again obviously in applications on Earth, although there do not appear to yet be associations that are exclusively dedicated to advocating for this form of energy and energy capture [37–43]. Although most existing scenarios are costly and only applicable on a small scale, this may change when any such technique or application is initiated on the Moon. In light of this, several sources that have been tested for their potential for energy harvesting may also be applicable in this new setting.

A partial list of potential areas for energy harvesting includes:

9.5.1 Floor pads in high-traffic areas

Energy harvesting devices have been tried in places such as major walkways or staircases in malls and have proven that energy can be harvested because people step in these areas most of the time during the mall's operating hours. Small piezoelectric sensors are embedded under the surface of the mat, and an AC voltage is generated each time a footstep pushes down on it. Existing sensors have been made largely of ceramics or crystal materials such as quartz and silicon dioxide in crystalline form.

This example is the first of three we examine that utilize the **piezoelectric effect**, the idea that the mechanical deformation of a material can generate some amount of electricity. Table 9.2 lists a small number of the many materials that exhibit this property, although there are many more, some of which have only been produced in tiny amounts. In broad categories, piezoelectric materials can be considered ceramic, polymeric, or synthetic. There can be overlap between the categories.

Table 9.2: Piezoelectric materials.

Formula	Name	Type	Comment
$BaTiO_3$	Barium titanate	Ceramic	
GaN	Gallium nitride	Ceramic	
$PbTiO_3$	Lead titanate	Ceramic	
$Pb(Zr_xTi_{1-x})O_3$	Lead zirconate titanate (PZT)	Ceramic	Synthetic
	Lead niobium zirconate titanate	Ceramic	Termed "soft PZT"
$(C_2H_2F_2)_n$	Polyvinylidene fluoride	Polymeric	
SiO_2	Quartz	Natural	Often laboratory grown
$Al_2SiO_4(F,OH)_2$	Topaz	Natural	
ZnO	Zinc oxide	Ceramic	Wide band gap

Figure 9.13 shows a basic schematic of how a piezoelectric material works. This deformation of the material, especially when it is in a high-traffic area within any lunar habitat, such as in a hallway or central area, means the cumulative generation of electricity can become significant over time.

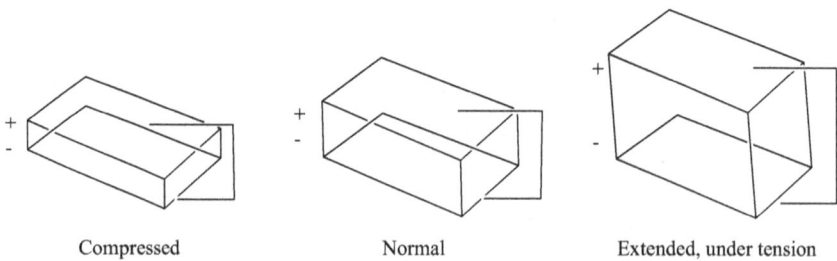

Compressed   Normal   Extended, under tension

Figure 9.13: Schematic of the piezoelectric effect.

9.5.2 Pads on stair steps

Like the just-mentioned floor pads, pads can be fitted to stairways, and each step can be a potential source of a small amount of energy. If installed selectively, these would be placed on stairways that are most heavily used. The previous figure, which illustrates how the piezoelectric effect can generate electricity on floor pads, will be the same type of device used here.

9.5.3 Boot heel pumps

Using the piezoelectric effect to gather energy from a person's walking may, again, seem like a technique that only yields a very small amount of energy. But the idea of what gets called piezoelectric footwear actually now has decades of research behind it. The idea is that:
1. Piezoelectric "pads" are placed under or within a shoe or boot sole, beneath the foot.
2. A diode is connected by solder to the piezoelectric pad.
3. The pads are linked to a battery pack [41].

If this were incorporated into footwear worn by all personnel living in a lunar colony, the potential amount of energy captured could be quite useful.

9.5.4 Revolving/sliding doors

There is a large potential on Earth for buildings with revolving doors to have such doors connected to some form of turbine, which, in turn, is connected to a generator. In a lunar habitat, it is more feasible to believe that numerous doors will be sliding, as they will require air-tight seals when closed. This requirement exists in case there is an unexpected pressure loss in any specific room – such doors can contain it and help limit the loss of pressure throughout a habitat. However, an advantage of sliding doors is that with each movement of the door, energy can be gathered simply by having a small wheel rolling against the door as part of a turbine.

9.5.5 Exercise equipment

Stationary bicycles, treadmills, stair-steppers
The idea of using a stationary bicycle to generate power is not a new one. Stationary bicycles hooked to some form of generator have been produced since at least the 1970s. Even further back, the idea of connecting a small rotating motor to a bicycle

tire, then to a light, is a decades-old means by which a bicycle can have a headlight, so it can be seen by others when it is ridden at night. Additionally, a stationary bicycle is part of the International Space Station as a means to keep astronauts healthy and to combat muscle weakness and bone loss. Since these devices already exist, they can be adapted to any stationary bicycles used as exercise equipment in a lunar colony.

The idea of running on a treadmill is also a very established one, with such treadmills being used extensively in gyms and individual domiciles. Currently, in essentially all cases of treadmill use, the energy supplied by the person running or walking on the treadmill is lost. However, the rolling bars at either end of the machine can be connected to generators, making this form of exercise equipment a source of harvestable energy. Figure 9.14 illustrates common treadmills. Figure 9.15 shows a common version of a stair-master, another exercise machine that has continuously moving parts which can be connected to small motors.

Figure 9.14: Treadmills.

Free weights, often used by serious exercise enthusiasts, may be a difficult source from which to obtain energy, but what are broadly termed weight machines are not. The repeated motion and thus the energy that goes into exercise using bench press machines, bicep curl machines, leg press or leg extension apparatuses, and the many other machines in existence can be gathered through the proper placement of small rotary motors or piston-like cylinders.

Clearly, the idea of energy harvesting in a very wide series of applications is one that needs a thorough examination in this novel, lunar setting.

Figure 9.15: Stair climber exercise machine.

References

[1]   American Solar Energy Society. Website. (Accessed 12 April 2025, as: https://ases.org).
[2]   Solar Energy Society. Website. (Accessed 12 April 2025, as: https://www.solarenergysociety.ca).
[3]   Solar Energy Society of Canada, Inc. (SESCI). Website. (Accessed 12 April 2025, as: https://resources. solarbusinesshub.com).
[4]   SolarPower Europe: Home. Website. (Accessed 12 April 2025, as: https://www.solarpowereu rope.org).
[5]   EUROSOLAR – The European Association for Renewable Energy. Website. (Accessed 12 April 2025, as: https://www.eurosolar.org).
[6]   International Solar Energy Society (ISES). Website. (Accessed 12 April 2025, as: https://www. ises.org).
[7]   Solar Energy UK. Website. (Accessed 12 April 2025, as: https://solarenergyuk.org).
[8]   Australian Solar Energy Society Ltd. Website. (Accessed 12 April 2025, as: https://www.acnc.gov.au).
[9]   Solar Energy Association of New Zealand. Website. (Accessed 12 April 2025, as: https://www.solaras sociation.org.nz).
[10] SESSA – Sustainable Energy Society of Southern Africa. Website. (Accessed 12 April 2025, as: https://sessa.org.za).
[11] SolarPACES. Website. (Accessed 12 April 2025, as: https://www.solarpaces.org).
[12] Aalborg CSP. Website. (Accessed 12 April 2025, as: https://www.aalborgcsp.com).

[13] Acciona. Website. (Accessed 12 April 2025, as: https://www.acciona.com).

[14] BrightSource Energy. Website. (Accessed 12 April 2025, as: https://www.brightsource.com).

[15] GlassPoint Solar. Website. (Accessed 12 April 2025, as: https://www.glasspoint.com).

[16] Novatec Solar. Website. (Accessed 12 April 2025, as: https://novatec.hr/en/services/renewable-energy-sources).

[17] Soliterm Group, GmbH. Website. (Accessed 12 April 2025, as: https://solitermgroup.com).

[18] Hitachi Energy. Website. (Accessed 12 April 2025, as: https://hitachienergy.com).

[19] Torresol Energy O&M S.L. – Madrid. Website. (Accessed 12 April 2025, as: https://www.torresolenergy.com).

[20] USGS Mineral Commodity Summaries, 2024. Downloadable as: https://doi.org/10.3133/mcs2024.

[21] Battery Council International. Website. (Accessed 12 April 2025, as: https://batterycouncil.org).

[22] The Rechargeable Battery Association. Website. (Accessed 12 April 2025, as: https://www.prba.org).

[23] Association of Battery Recyclers. Website. (Accessed 12 April 2025, as: https://www.associationofbatteryrecyclers.com).

[24] The United States Advanced Battery Consortium LLC (USABC). Website. (Accessed 12 April 2025, as: https://uscar.org/usabc/).

[25] Canadian Battery Association. Website. (Accessed 12 April 2025, as: https://canadianbatteryassociation.ca).

[26] BMAC – Battery Metals Association of Canada. Website. (Accessed 12 April 2025, as: https://www.bmacanada.org).

[27] The British and Irish Portable Battery Association. Website. (Accessed 12 April 2025, as: https://bipba.co.uk).

[28] European Battery Alliance. Website. (Accessed 12 April 2025, as: https://www.eba250.com).

[29] EUROBAT. Website. (Accessed 12 April 2025, as: https://www.europbat.org).

[30] Batteries Europe. Website. (Accessed 12 April 2025, as: https://batterieseurope.eu).

[31] Australian Battery Society. Website. (Accessed 12 April 2025, as: https://australianbatterysociety.org).

[32] Battery Association of Japan. Website. (Accessed 12 April 2025, as: https://www.baj.or.jp).

[33] Korea Battery Industry Association. Website. (Accessed 12 April 2025, as: https://naatbatt.org).

[34] Indian Battery Manufacturers Association. Website. (Accessed 12 April 2025, as: https://ibma.org.in).

[35] South African Energy Storage Association. Website. (Accessed 12 April 2025, as: https://saesa.org.za).

[36] Eco Renewable Energy – Innovations That Energise. Website. (Accessed 12 April 2025, as: https://www.ecorenewableenergy.com.au/).

[37] Solar Impulse Foundation – Off the Grid Bikes. Website. (Accessed 12 April 2025, as: https://solarimpulse.com/solutions-explorer/off-the-grid-electricity-generating-spinning-bike#:~:text=Off%20The%20Grid%20spinning%20bikes,into%20the%20building's%20electrical%20grid).

[38] Eco renewable energy. Get Fit and Generate Energy From These Innovative Electricity Generating Bikes. Website. (Accessed 12 April 2025, as: https://www.ecorenewableenergy.com.au/articles/change-the-world-one-workout-at-a-time-eco-powr-spin-bikes/#:~:text=The%20bike%20houses%20an%20integrated%20micro%2Dinverter%20%E2%80%93,power%20grid%2C%20thus%20offsetting%20the%20power%20consumption.&text=Eco%2DPowr%20spin%20bikes%20are%20eco%2Dfriendly%20because%20they,is%20a%20renewable%20and%20sustainable%20energy%20source).

[39] ASME Energy Harvesting Technical Committee. Website. (Accessed 12 April 2025, as: https://www.asme.org/get-involved/groups-sections-and-technical-divisions/technical-divisions/technical-divisions-community-pages/smasis-division/technical-committees-(1)/energy-harvesting#:~:text=Description,academic%20disciplines%20and%20industrial%20sectors).

[40] European Sustainable Energy Innovation Alliance (ESEIA) and the Graphene Flagship's GRAPHERGIA project. Website. (Accessed 12 April 2025, as: https://graphene-flagship.eu/#:~:text=The%20GRAPHE RGIA%20project%20has%20launched,or%20advancing%20future%20battery%20technologies).

[41] B. Zhao, F. Qian, A. Hatfield, L. Zuo, and T.-B. Xu. A review of piezoelectric footwear energy harvesters: Principles, methods, and applications, *Sensors*, 23:13. doi: 10.3390/s23135841.

[42] PSMA, Energy Harvesting Forum. Website. (Accessed 12 April 2025 as: https://psma.com/technical-forums/energy-harvesting).

[43] American Society of Mechanical Engineers. Energy Harvesting. Website. (Accessed 12 April 2025 as: https://www.asme.org/get-involved/groups-sections-and-technical-divisions/technical-divisions/tech nical-divisions-community-pages).

Index

https://doi.org/10.1515/9783111387307-010

www.ingramcontent.com/pod-product-compliance
Lightning Source LLC
Chambersburg PA
CBHW081552220326
41598CB00036B/6652